RELIABILITY OF
SHELL BUCKLING PREDICTIONS

RELIABILITY OF SHELL BUCKLING PREDICTIONS

WILLIAM A. LITLE

RESEARCH MONOGRAPH NO. 25
THE M.I.T. PRESS, CAMBRIDGE, MASSACHUSETTS

Library of Congress Catalog Card Number: 64-19394
Printed in the United States of America

FOREWORD

There has long been a need in science and engineering for systematic publication of research studies larger in scope than a journal article but less ambitious than a finished book. Much valuable work of this kind is now published only in a semiprivate way, perhaps as a laboratory report, and so may not find its proper place in the literature of the field. The present contribution is the twenty-fifth of the M.I.T. Press Research Monographs, which we hope will make selected timely and important research studies readily accessible to libraries and to the independent worker.

J. A. Stratton

PREFACE

The Laboratory for Structural Models of M.I.T.'s Department of Civil Engineering has, since 1960, been increasingly concerned with the development and usage of the structural model as an educational, a design, or a research instrument. This development is based on the conviction that the solution of many structural design problems is impeded by the lack of a thorough, well-understood experimental design process. This monograph represents one aspect of the work that is proceeding in the Laboratory.

It should be pointed out that all of the work is directed toward the solution of civil engineering problems, which most often pertain to one-of-a-kind situations. The author attaches particular significance to the oneness concept, and feels that many (but by no means all) experimental techniques must, for practical reasons, be evaluated in light of potential area of application.

The Laboratory is directed by Dr. R. J. Hansen. His encouragement and support, as well as that of other members of the Laboratory staff, are gratefully acknowledged.

Cambridge, Massachusetts W. A. Litle
May, 1964

CONTENTS

SUMMARY

OBJECT: The experimental method of structural design has a unique and obvious advantage over the mathematical method in those problems wherein the structural behavior is not well understood. One such problem is that of the buckling of thin-shell roof structures. The purpose of this monograph is to examine the reliability of small-scale plastic models in the determination of elastic buckling pressures of thin-shell structures.

SCOPE: Analytical work is presented to show how and to what extent the theories of probability and statistics can be applied in the experimental design method. An experimental program is intended to reveal the consistency of plastic material properties, the repeatability of buckling pressures from different models, the effects of thickness and middle surface geometry variation, the effect of geometric scale, the effect of boundary conditions, and the effect of the means of applying load.

PROCEDURE: The spherical cap is one of the most thoroughly investigated thin-shell structures. Because of the degree of comparison and control which these previous investigations could afford, the spherical cap was chosen for the present study. A Type I polyvinyl chloride plastic was used as the model material. Its elastic properties were thoroughly investigated. Twenty 18-inch radius domes were fabricated from six different, but supposedly identical, molds. All of these were subjected to air pressure loading and four of them were loaded with weights. Four 36-inch radius domes were fabricated from two different molds which were supposedly twice the size of the 18-inch radius molds. The 36-inch domes were loaded only with air pressure.

CONCLUSIONS: 1. Experimental results obtained from

structural models can be used to predict something about the average or mean value of a particular physical quantity in a prototype structure. On the other hand, it is in general not possible to determine correct results regarding the dispersion about this derived mean value.

2. If the model result is taken to be the result obtained from a single model, then it is particularly important that the investigator be convinced of the absence of blunders and major systematic errors.
3. The elastic properties of the polyvinyl chloride plastic will not vary widely throughout a vacuum-formed shell model and can be determined within 5%.
4. Average shell thickness should be taken as the controlling thickness. In this light, minor (±10%) thickness variations do not significantly affect the buckling pressures.
5. Buckling behavior is mildly affected by middle surface geometric variations. Although only a narrow thickness range was studied, the experimental results showed that buckling pressures were proportional to $(t/R)^{2.9}$ instead of $(t/R)^2$ as predicted by theory. How much of this discrepancy is due to geometric imperfections is not known. In any case, the imperfection effect on the imaginary prototype structure that is being designed cannot be deduced, and therefore, the extrapolation from model to prototype must incorporate a safety factor to allow for the effect of imperfection.
6. Buckling behavior is extremely sensitive to changes in boundary restraint. If the model loading rig induces initial edge bending, the buckling behavior will almost certainly be affected. Any buckling model must reproduce as nearly as possible the prototype boundary condition, and if the prototype condition is uncertain, then provision should be made for alternate model tests in which the possible prototype conditions are bracketed.
7. The technique of applying load by hanging weights on a grid pattern leads to the same buckling pressures as would be obtained with a continuous load. In fact, pressure loading is not recommended unless the shell is very flat.

This is because the nature of the buckling action seems to be dependent upon load direction as well as stress level.

8. An epoxy-cemented edge apparently did not induce any initial bending. For all tests made with such edge restraint, the variation in air buckling pressure never exceeded 20%. For engineering design, such reliability is considered to be very good, and it is concluded that the sensitivity of a thin shell to buckling is not so great as to preclude useful structural model design.
9. The buckling behavior of the model shells was not affected by a 100% change in the geometric scale.

Chapter 1

INTRODUCTION

The structural consequences of any building design may be evaluated by the grace and efficiency with which the forces of nature are resisted. The class of structures which Nervi refers to as "form resistant" are perhaps the most efficient and graceful of all. They are efficient since their geometric form allows them to carry load with little or no bending, and they are graceful because the magnitude and combination of curvature can be selected in such a way that the structural unit conveys an aesthetically pleasing impression. Consequently, it is of interest to consider why such construction is relatively uncommon. Perhaps there are three principal reasons. First, our understanding of the manner in which these structures resist forces is extremely inadequate. The mathematical equations for even the most simple situations are often extremely cumbersome if, in fact, they can be dealt with at all. Second, it must be noted that over-all economy cannot be measured by minimum weight alone and, hence, one must consider the total construction cost. The need to prepare formwork for a reinforced concrete thin-shell roof may add 50 to 100% to the cost of the structure depending upon the complexity of the surface. Thus because of a temporary construction situation a designer may decide to utilize another type of construction. Finally, it must be understood that there are many situations in which the functional requirements for the building will not permit the use of shell construction. In fact, one might seriously wonder whether this third reason—even assuming that the analysis and formwork problems were completely solved—might, by itself, preclude a rapid increase in the use of "form resistant" construction.

There are, of course, many aspects to the problem of understanding the structural behavior of thin-shell or "form resistant" constructions. This monograph is concerned only with the aspect of instability.

1.1 The Thin-Shell Structural Design Process

Once the geometric form and type of construction have been selected for any building, the only demand on the structure is that it maintain in equilibrium all of the applied forces without deforming so much as to interfere with the function of the building. Thus, the structure must simply possess sufficient strength and sufficient stiffness to resist the environmental conditions to which it will be subjected. The engineer has available certain "tools" which can aid him in establishing the sufficiency. Generally speaking, these "tools" could be either of a mathematical or an experimental nature.

Throughout the course of history it has been true that the pioneering structures were designed and constructed without the aid of applicable mathematical theory. Then, with the passage of time, man would strive to understand better the behavior of the structures until, finally, he would be able to describe the behavior in mathematical terms. And so it is today; simple beams long ago fell to the power of the applied mathematician; the behavior of statically indeterminate planar frameworks has more recently been solved; and the category of "form resistant" structures is not yet well understood. Doubtless the day will come when this latter type of structure will yield to the mathematical analysis, but in the meantime such structures will continue to be built. It is just this sort of situation in which the experimental design process can play a significant part. For in those cases where the structural behavior cannot be described mathematically, a carefully constructed and tested scale model should be able to reveal it. Of course, it should be realized that there are different types of physical models and that they are intended to satisfy different objectives. The architects have for many years made considerable use of the visual model, often constructed of cardboard, wood sticks, etc. Such models <u>may</u> also permit one to ascertain certain qualitative information regarding the behavior of the structural system. By distinction, physical models can be manufactured and tested which will yield useful quantitative information—be it about mode of failure, magnitude of stress, magnitude of failure load, etc. It is this latter, quantitative, model that is of concern here.

1.2 History of Previous Shell Stability Studies

Before entering into the main body of this monograph, it should be useful to pause for a minute and outline briefly the history and present status of the mathematical and the experimental approach to shell instability. To have a clear picture of the present state of each approach is of inestimable value, because in any actual problem the designer should bring to bear all of the available resources. Nash[30, 31]* has compiled a bibliography on shell structures and lists 1455 books and papers prior to 1954 and 884 between 1954 and 1956 alone. Certainly many, many more have been written since, and so it would be completely unreasonable

*Superscript numbers refer to entries in the References.

to make any kind of complete survey here.

Fairbairn's[12] study in 1858 of the buckling of cylinders under external pressure was one of the first experimental investigations of shell stability. Stability theory goes back to Euler and Lagrange, but Bryan[7] (1888) gave perhaps the first general discussion of shell stability. In line with the bifurcation concept of buckling, he suggested that the buckling load could be determined from the fact that there would be no change in the total potential energy as the shell passed from the unbuckled to the buckled state.

In this century a tremendous effort has been expended on the shell stability problem. At first, the theoretical workers pursued the approach which had been experimentally verified for the column and flat-plate buckling problems. Thus, they composed nonlinear equilibrium equations; but then, considering that the deflections would be small, proceeded to linearize the equations by assuming that the membrane forces in the buckled configuration would be the same as they had been in the unbuckled state. Southwell[41] (1914) and others were concerned with cylindrical tubes, and Zoelly[47] (1915) concerned himself with the complete sphere. Naturally, experiments were conducted in an attempt to verify the theoretical developments. The results of these experimental studies were startling. In some cases, such as the cylindrical tube under external pressure, the agreement between theory and experiment was satisfactory. On the other hand, the equally simple problems of the cylindrical tube under axial load and the radially loaded sphere led to a complete dichotomy.

The set of linearized differential equations for the cylindrical tube subjected to a general type of loading as given by Flügge[14] are

$$u'' + \frac{1-\nu}{2}u^{\cdot\cdot} + \frac{1+\nu}{2}v'^{\cdot} + \nu w' + \frac{t^2}{12R^2}\left(\frac{1-\nu}{2}u^{\cdot\cdot} - w''' + \frac{1-\nu}{2}w'^{\cdot\cdot}\right)$$

$$- \frac{pR(1-\nu^2)}{Et}(u^{\cdot\cdot} - w') - \frac{P(1-\nu^2)}{Et}u'' - \frac{2T(1-\nu^2)}{Et}u'^{\cdot} = 0 \tag{1.1a}$$

$$\frac{1+\nu}{2}u'^{\cdot} + V^{\cdot\cdot} + \frac{1-\nu}{2}v'' + w^{\cdot} + \frac{t^2}{12R^2}\left[\frac{3}{2}(1-\nu)v'' - \frac{3-\nu}{2}w''^{\cdot}\right]$$

$$- \frac{pR(1-\nu^2)}{Et}(v^{\cdot\cdot} + w^{\cdot}) - \frac{P(1-\nu^2)}{Et}v'' - \frac{2T(1-\nu^2)}{Et}(v'^{\cdot} + w') = 0 \tag{1.1b}$$

$$
\begin{aligned}
&\nu u' + v^{\cdot} + w \\
&+ \frac{t^2}{12R^2}\left(\frac{1-\nu}{2}u'^{\cdot\cdot} - u''' - \frac{3-\nu}{2}v''^{\cdot} + w'''' + 2w''^{\cdot\cdot} + w^{\cdot\cdot\cdot\cdot} + 2w^{\cdot\cdot} + w\right) \\
&+ \frac{pR(1-\nu^2)}{Et}(u' - v^{\cdot} + w^{\cdot\cdot}) + \frac{P(1-\nu^2)}{Et}w'' - \frac{2T(1-\nu^2)}{Et}(v' - w'^{\cdot}) = 0
\end{aligned}
\tag{1.1c}
$$

where the notations are explained in Figure 1.1.

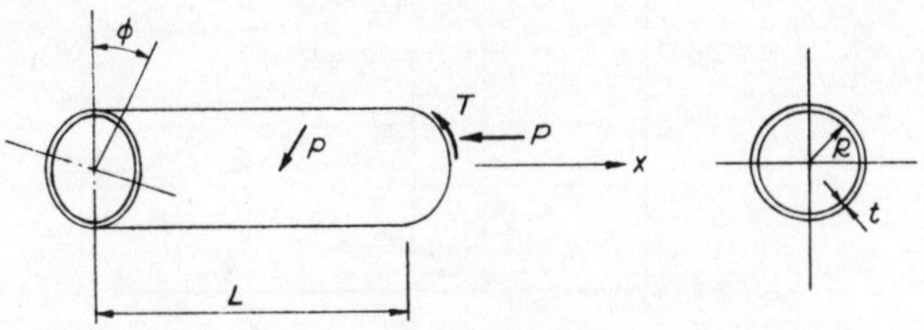

u = Displacement Along Generator, + in Direction of Increasing x

v = Displacement Along a Circle of Radius R + z, + in Direction of Increasing φ

w = Radial Displacement, + Outward

$(\ )' = R\frac{\partial(\ )}{\partial x}$

$(\ )^{\cdot} = \frac{\partial(\ )}{\partial \phi}$

p = Radial Pressure

P = Axial Force / Unit Length Along Circumference

T = Shear Force / Unit Length Along Circumference

Figure 1.1. Notation for cylindrical shell theory.

The results from a solution of these equations for the case of a tube with simply supported ends subjected only to external pressure are shown in Figure 1.2. Experimental results obtained by Sturm[42] are also shown, and it is seen that the agreement between theory and experiment is good.

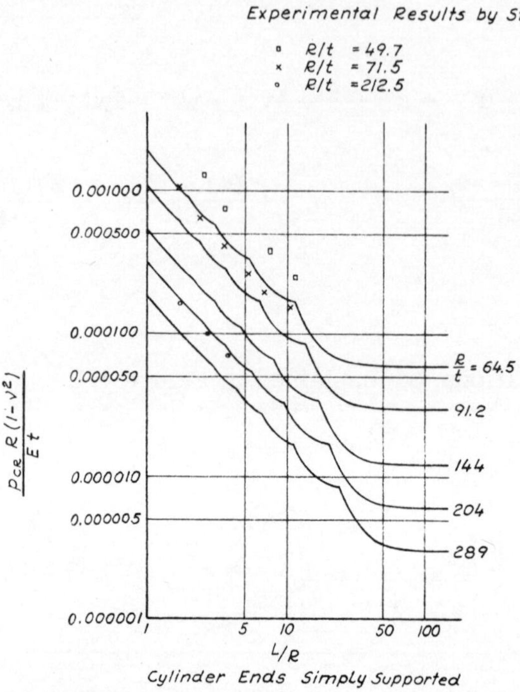

Figure 1.2. Elastic buckling of cylindrical shells under external pressure.[14]

For the case of axial compression, the set of Equations 1.1 can again be specialized and solved. The type of behavior to be expected is shown in Figure 1.3. For very short cylinders the behavior is analogous to Euler buckling of a strip element, and for very long cylinders the tube acts as an Euler column. For intermediate lengths the solution tends toward the result which corresponds to a symmetrical rippling of the side walls. The critical stress for this case is given by

$$\sigma_{\text{critical}} = \frac{P}{t} = \frac{E}{\sqrt{3(1-\nu^2)}}\left(\frac{t}{R}\right) \tag{1.2}$$

When experiments were made to check the validity of these results, a behavior was observed which was completely at variance with theory. Figure 1.4 shows the results of Lundquist[26] and Donnell.[9] It is apparent that experiments indicate critical stress levels on the order of 1/3 those given by the classical linear theory.

In searching for an answer as to why there was such a discrepancy between experiment and theory, Donnell[9] (1934)

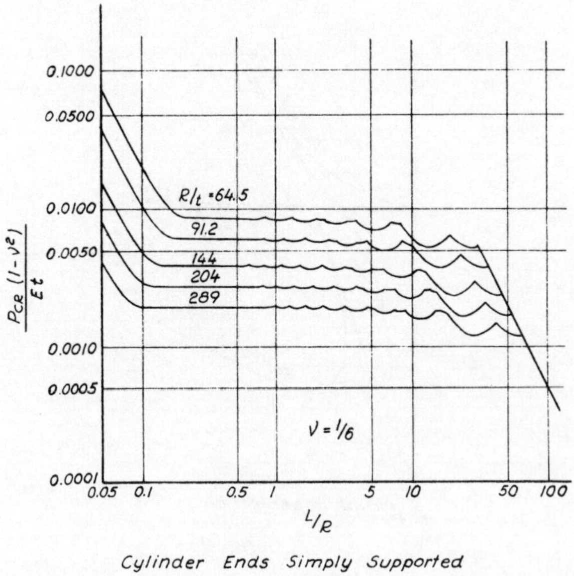

Figure 1.3. Elastic buckling of cylindrical shells under axial compression.[14]

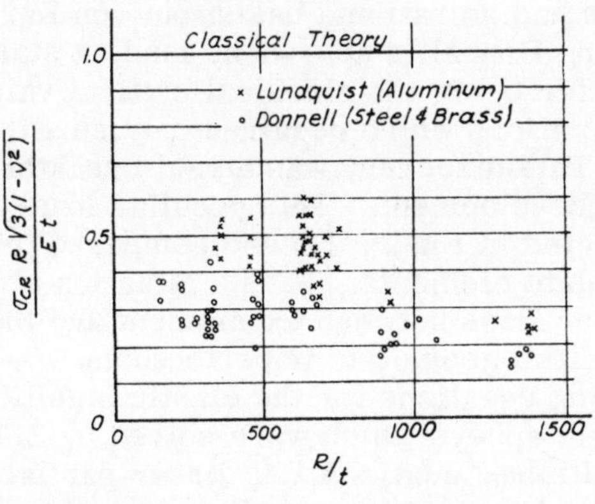

Figure 1.4. Discrepancy between experiment and theory for axially loaded cylindrical shells.

decided to track the load-deflection relationship of the shell. By starting the solution with an assumed imperfection and using nonlinear strain-displacement relationships of the form

$$\epsilon_x = \frac{\partial u}{\partial x} + \frac{1}{2}\left(\frac{\partial w}{2x}\right)^2 \tag{1.3}$$

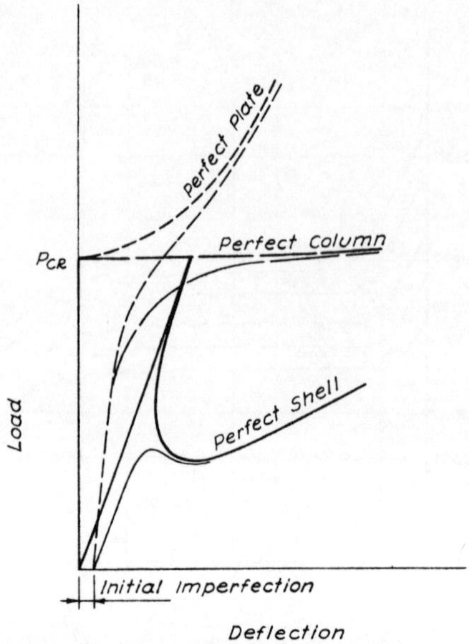

Figure 1.5. Stability behavior of different geometric forms.

small but finite displacements could be treated. Donnell[9] imposed a first yield criterion as the limit capacity. The first yield approach had no rational basis; but von Karman and Tsien[22] (1939), utilizing Donnell's approach, made a startling discovery that for loads far below the classical critical value there were equilibrium positions which required only small, but finite, displacements. This discovery was a real breakthrough in the mathematical development. This peculiar load-deflection behavior is depicted in Figure 1.5 and compared to the behavior of columns and plate elements. Such a situation almost begs one to explain differences between experiment and theory by taking into account initial geometric imperfections.

The governing equations for the elastic stability of the radially loaded complete sphere which were solved by Zoelly[47] are, like the cylindrical tube Equations 1.1, linear partial differential equations. As given by Flügge[14] these are

$$\left(1+\frac{t^2}{12R^2}\right)[v^{\cdot\cdot}+v^{\cdot}\cot\phi - v(\nu+\cot^2\phi)] + (1+\nu)w^{\cdot}$$

$$-\frac{t^2}{12R^2}[w^{\cdot\cdot\cdot}+w^{\cdot\cdot}\cot\phi - w^{\cdot}(\nu+\cot^2\phi] - \frac{pR(1-\nu^2)}{2Et}(v-w^{\cdot}) = 0$$

(1.4a)

$$(1+\nu)(v^{\cdot} + v\cot\phi + 2w)$$

$$+ \frac{t^2}{12R^2}[- v^{\cdot\cdot\cdot} - 2v^{\cdot\cdot}\cot\phi + v^{\cdot}(1+\nu+\cot^2\phi) - v(2-\nu+\cot^2\phi)\cot\phi$$

$$+ w^{\cdot\cdot\cdot\cdot} + 2w^{\cdot\cdot\cdot}\cot\phi - w^{\cdot\cdot}(1+\nu+\cot^2\phi) + w^{\cdot}(2-\nu+\cot^2\phi)\cot\phi]$$

$$+ \frac{pR(1-\nu^2)}{2Et}(v^{\cdot} + v\cot\phi + w^{\cdot\cdot} + w^{\cdot}\cot\phi + 4w) = 0 \qquad (1.4b)$$

where the notations are explained in Figure 1.6.

The solution of these two linear partial differential equations yields for the lowest critical pressure

$$p_{cr} = \frac{2E}{\sqrt{3(1-\nu^2)}}\left(\frac{t}{R}\right)^2 \qquad (1.5)$$

For the spherical shell loaded by radial pressure, the membrane stress is everywhere equal to

$$\sigma = \frac{pR}{2t} \qquad (1.6)$$

so that the critical stress is

$$\sigma_{cr} = \frac{E}{\sqrt{3(1-\nu^2)}}\left(\frac{t}{R}\right) \qquad (1.7)$$

the same as for the axially loaded cylinder. Like the axially loaded cylinder, experiments consistently yielded pressures of

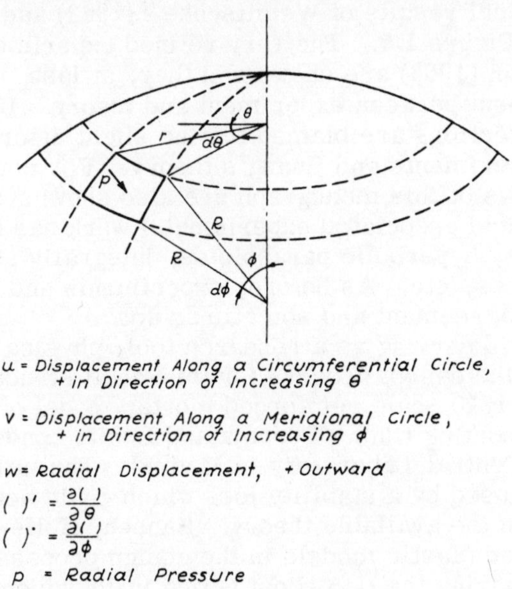

Figure 1.6. Notation for spherical shell theory.

1/4 to 1/2 of the theoretical value. In Figure 1.7 the results of a large number of experiments on spherical caps are compared to this theoretical value.

Again, the theoreticians turned to the approach of tracking the load-deflection relationship through the solution of nonlinear partial differential equations. Further, they considered the spherical cap rather than the complete sphere. Usually the edge of the spherical cap was assumed fixed, so that the shell did not remain spherical up to the critical pressure. In this way, the shell could introduce its own initial imperfection during loading, even though it was a perfect sphere at the start of loading. To simplify the mathematics, they restricted themselves to shallow spherical caps (rise less than about 1/16 base diameter) undergoing axisymmetric deflection. Thus by starting with one of the simplest of shell problems and assuming the critical load to be that of the first maximum on the load-deflection curve, a massive effort has yielded the result shown in Figure 1.8. These theoretical results, while identifying the opening angle as an additional parameter, still fail to explain the low experimental pressures. (It should be noted that except for the tests reported in this monograph, and those of Klöppel and Jungbluth, all of the data in Figure 1.7 are for shallow shells.) Von Karman and Tsien (1939) proposed an energy jump buckling criterion, but this proposal calls for a changeable result depending upon the stiffness characteristics of the loading mechanism. Tests do not show any such variation. The latest thoughts are that the assumption of axisymmetric buckling overlooks other asymmetric buckling modes which correspond to lower pressures.

The analytical results of Weinitschke[45] (1962) and Huang (1963) are given in Figure 1.9. The very refined experiments of Krenzke and Kiernan (1963) are shown and they, at last, indicate a close agreement between experiment and theory. Unavoidable initial imperfections are blamed for the slight discrepancy between the experiments and Huang's theory. For comparison, the test results of this monograph are also shown.

Analytical and associated experimental work has also been done on cones, hyperbolic paraboloids, integrally stiffened shells of various types, etc. As before, experiments and theory sometimes are in agreement and sometimes not.

In addition to serving as a research tool, physical models have been used in the design process for thin-shell structures. Figure 1.10 shows a 1/10 scale reinforced mortar model of the Tachira, Venezuela, Sporting Club. Torroja and Benito conducted this study at the Central Laboratory in Madrid. The failure of the model was caused by a stability loss which could not have been predicted with the available theory. Hansen, Holley, and Biggs at M. I. T. used plastic models in the design process for the thin-shell roof covering the U. S. Post Office in Providence, Rhode Island. The stability problem was completely solved using 1/80

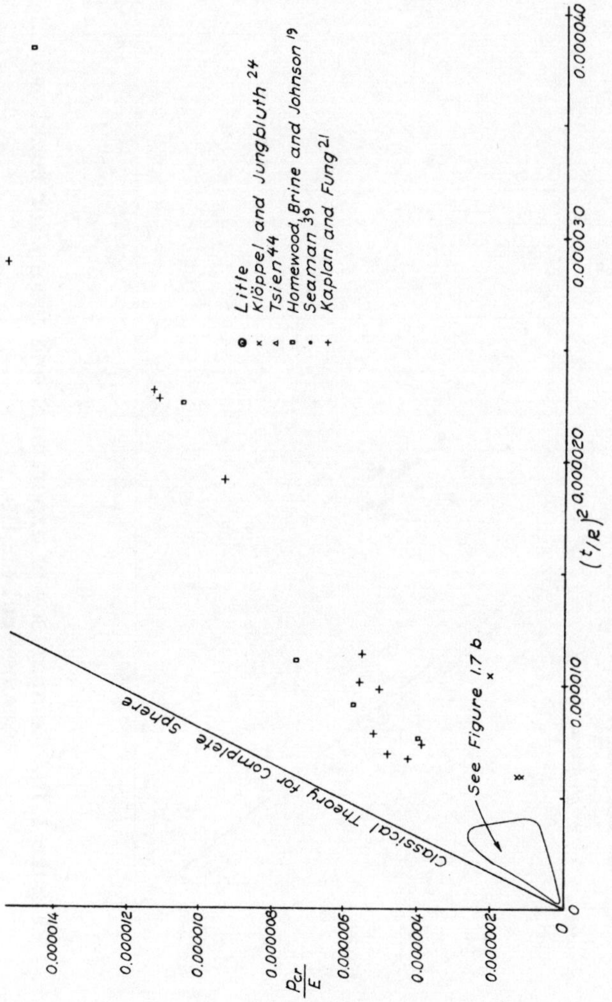

Figure 1.7a. Comparison of experiment and theory for buckling of spherical shells.

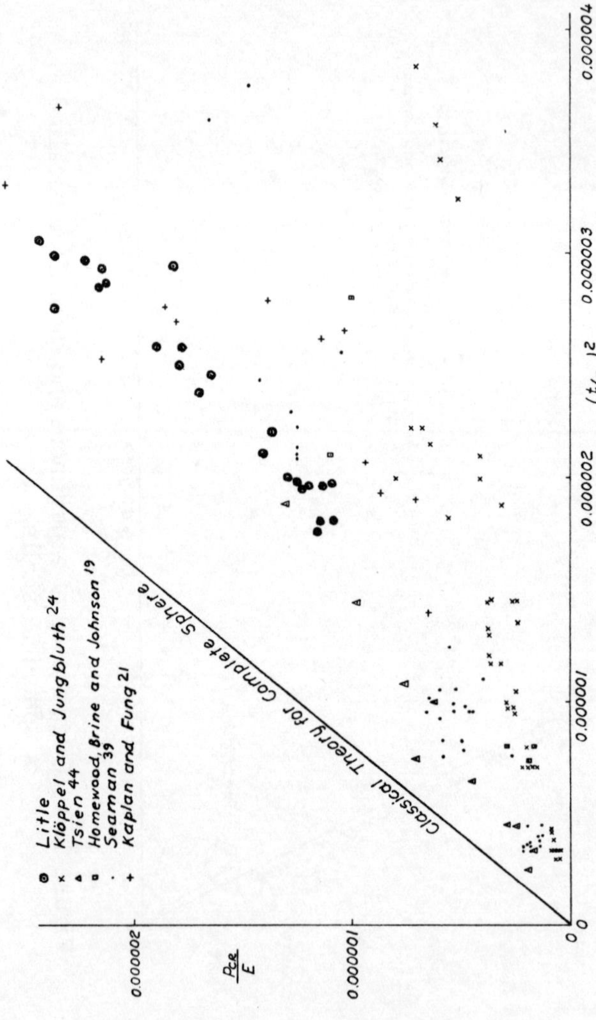

Figure 1.7b. Comparison of experiment and theory for buckling of spherical shells.

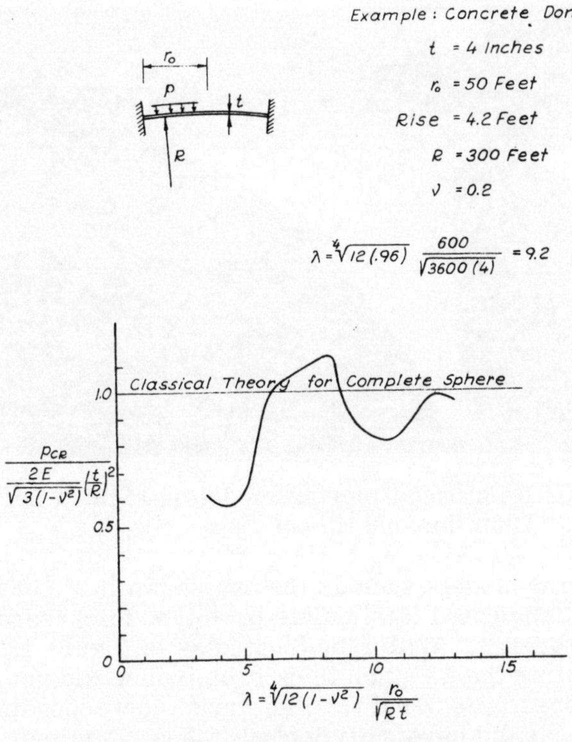

Figure 1.8. Calculated buckling pressures of initially perfect clamped shallow spherical shells.

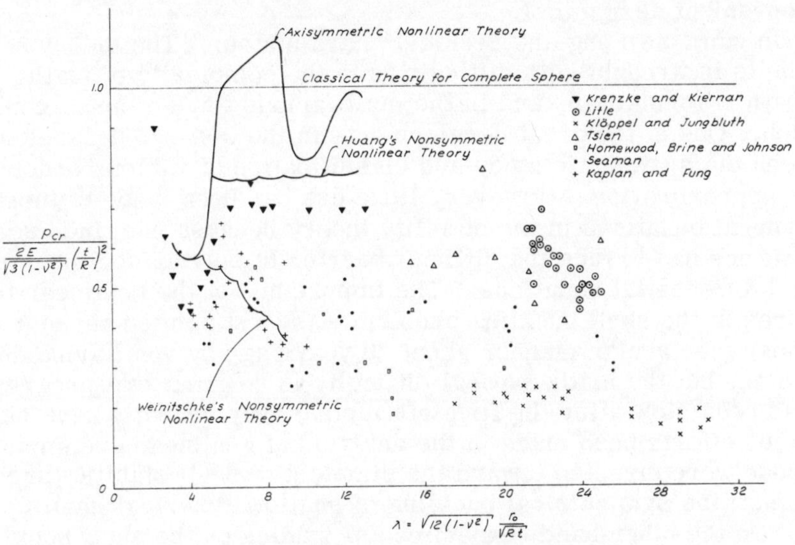

Figure 1.9. Axisymmetric and nonsymmetric theories compared to experimental results.

Figure 1.10. Reinforced mortar model of Tachira Sporting Club thin-shell roof.

scale structural models such as the one shown in Figure 1.11. Models were fabricated to simulate prototype thicknesses of $4\frac{1}{2}$, 6, $7\frac{1}{2}$, and 9 inches. With ribs located as in Figure 1.11, the model simulating the $4\frac{1}{2}$-inch thick prototype would not carry the required load. It was found that the model corresponding to the 6-inch thickness did have sufficient stiffness. Minimum rib dimensions were then determined by testing, altering, retesting, etc., the same model. Like the Tachira Club shell, mathematical stability theory was nonexistent; but the model was a very convenient design tool.

In sum, as Fung and Sechler[15] have stated: "The theory of shells is wrought with difficulties in the complexity of mathematical formalism, and in the multifarious ways of approximation. One of the great developments in the last two decades has been the systematic study and classification of various orders of approximation. However, little use has been made of these general equations in the stability theory because one finds soon that one has to face the difficult barrier of solving nonlinear partial differential equations. The importance of the nonlinear features in the shell buckling problem was first pointed out in a most spectacular manner about 20 years ago by von Karman and Tsien, but the mathematical difficulty is so great that progress has been slow after the first attempts. Only recently have serious efforts been made in the analysis of nonlinear systems, and these were directed toward the simplest of shell stability problems: the symmetrical buckling of shallow spherical shells.

"On the other hand, experimental studies on the shell buckling problem are also encompassed with difficulties. The wide scatter

Figure 1.11. Buckling models for Providence Post Office.

and the nonrepeatability of experimental results attest to the experimental difficulty and to the demand for careful attention to testing methods."

1.3 Object and Scope

The purpose of this monograph is to study the reliability of the information which a small-scale physical model will yield in the area of the elastic instability of thin-shell roof structures. Material is presented that serves to show how and to what extent statistics and probability theory can be applied in model programs. An experimental program is carried out to deduce the reliability of thin-shell polyvinyl chloride plastic models. In so doing it is intended to consider such things as repeatability, consistency of material properties, effects of thickness and middle surface geometry variation, boundary conditions, scale effect, and the means of applying load.

The problem is approached from the point of view of attempting to develop the experimental design process, not to aid in the development of mathematical theory. It is hoped, of course, that the results will also be of interest to the theoretician.

Chapter 2

THE MODEL PROCESS FOR ELASTIC STABILITY OF SHELL STRUCTURES

The complete process of any structural model analysis falls naturally into five parts, namely: (1) planning the study, (2) model fabrication, (3) loading, (4) testing and recording information, and (5) interpreting the recorded data and extrapolating them to the prototype. These five phases are rather closely interrelated. Consequently it is mandatory that, prior to the initiation of any project, consideration be given to exactly how each step of the process is to be accomplished and that each step is compatible with all others. The intention in this chapter is to set down the factors that were considered at the outset of the project.

2.1 Planning the Study

In an actual design study the results obtained from the model itself must be interpreted or operated upon in light of the prototype; for, after all, it is the prototype result, not the model result, which is of interest. The restrictions of time and cost will usually limit the number of model tests which can be carried out. Previous experimental results have shown a wide scatter. If, then, one were to make only a very few model tests, what could be said of the extrapolated prototype result? This question involves many factors. First of all, any model testing procedure and the results obtained therefrom will incorporate a variety of errors. Further, the constitution (that is, geometry, material properties, etc.) of the prototype and model are not certain in a probabilistic sense. In such a light, one could ask whether, with test data from a limited number of model tests, it would be possible to state with a certain confidence level the range within which the still imaginary prototype result would fall. An error analysis study presented in Chapter 3 shows that some mathematical foundations exist for treating random errors. On the other hand, it shows that if the experiment is dominated by sources of systematic error, then the mathematical methods of statistics and probability are of no use.

The wide scatter in some of the previous shell buckling tests raised some question whether the shell instability problem was so sensitive to sources of random and systematic error as to make experimental design an unreliable process. Several questions were raised. Is the internal constitution (that is, geometry, material properties, etc.) of a shell so critical that two adjacent models fabricated and tested in the same manner will yield widely separated results? Will two adjacent models, fabricated and tested in the same manner but having a different geometric scale, yield the same result? How sensitive is the model to variations in boundary conditions? Since any model will have some thickness variation, what is the thickness which should be assigned to the model when it is compared to the prototype? Do model thickness variations crucially affect the buckling pressures? How accurately must the geometric surface in the model be controlled? Does the means of loading have a significant effect on the buckling pressure? All previous experimental programs have involved the variation of some parameters such as the radius to

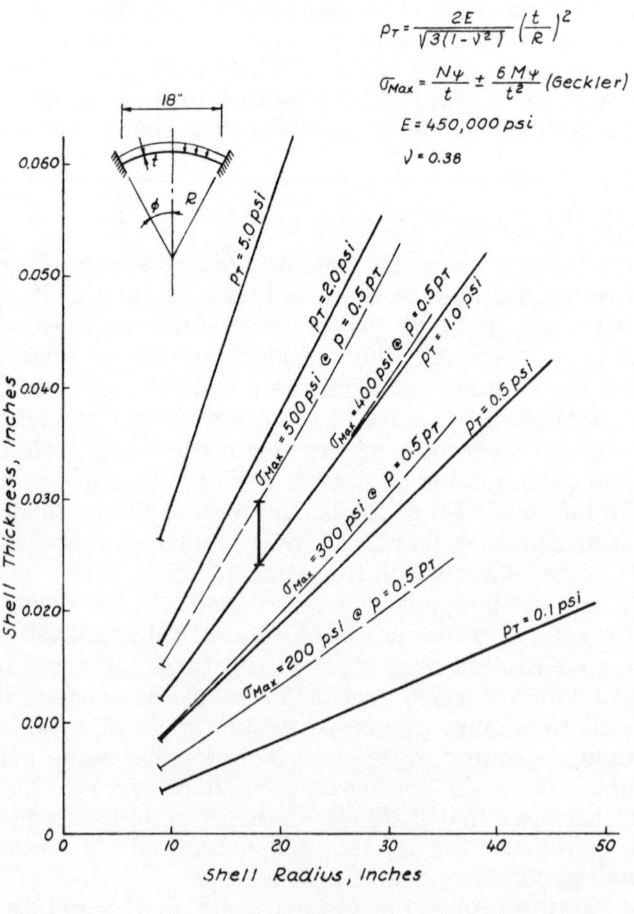

Figure 2.1. Some consequences of the R/t selection.

thickness ratio that are pertinent to the theoretical solution. To the author's knowledge no experimental programs have attempted to answer any of these questions—questions that must be answered if the experimental method is to be used in the shell design process.

The spherical cap was chosen as the structure to be modeled. The principal reason for selecting this shape is also a factor which makes the choice an unfortunate one. Thus, the fact that a great theoretical and experimental effort had already been expended on this shape allowed for a measure of control and comparison; however, a shape such as that used for the Tachira Club shown in Figure 1.10 demonstrates much more effectively the power of the model approach.

Previous experience in the laboratory had demonstrated the usefulness of plastics as model materials, and in fact special fabricating equipment had been installed which permitted rapid

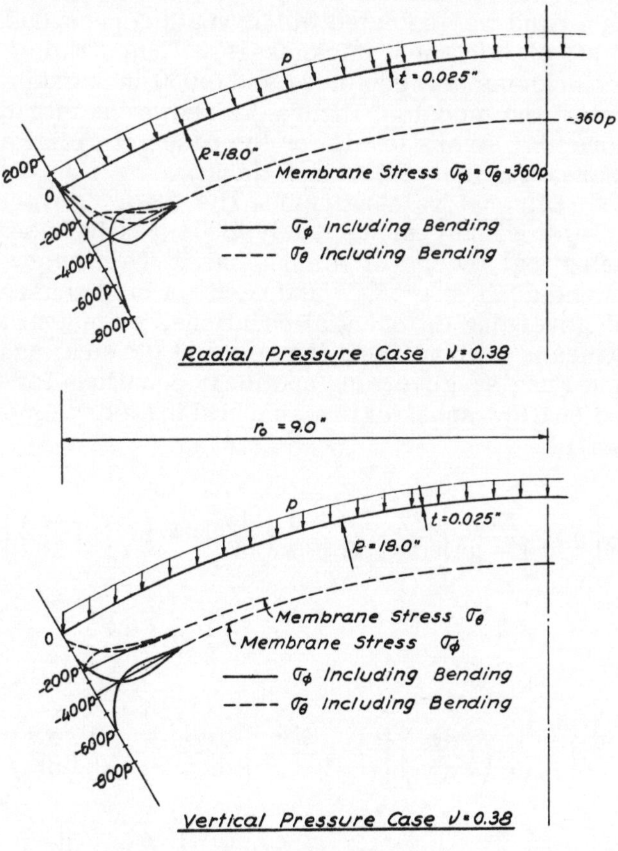

Figure 2.2. Stress distributions in model shell.

fabrication of plastic shell models with a plan area of up to 20 by 20 inches. Figure 2.1 depicts some of the consequences of the selection of a radius to thickness ratio for the model spheres, given a base diameter of 18 inches, a modulus of elasticity of 450,000 psi, and Poisson's ratio = 0.38. It was planned to use air pressure for loading, and a simple water manometer would show little error if the buckling pressures exceeded 0.5 psi. Allowing for the previous discrepancy between theory and experiment, it was decided to restrict the radius to thickness ratio to those values such that the critical theoretical buckling pressure for the complete sphere was greater than 1.0 psi. Also to be considered was the fact that stress levels could not be high; otherwise the behavior would no longer be elastic. Using the Geckler approximation, the maximum stress (at the fixed edge) was computed for a pressure equal to 1/2 the theoretical buckling pressure. Keeping the maximum stress below 500 psi was satisfactory. From the range of now permissible radius to thickness ratios a band was selected which would correspond to readily available plastic sheet thicknesses. Thus a radius of 18 inches and a thickness between 0.024 and 0.030 inch established the geometry of the models. Figure 2.2 shows the meridional and circumferential stress levels for the cases of radial and vertical pressure on a shell 0.025 inch thick.

2.1.1 Dimensional Analysis. The theory governing and tying together any model analysis is really based upon the mathematical theory of dimensions. The theory is well-documented[3,6,8,17,25,29,34] and need not be discussed here.

The governing differential equations, when available, are an ideal means of identifying the relevant physical quantities. Kaplan and Fung's[21] governing nonlinear equations for the radially loaded shallow spherical cap undergoing axisymmetric deformations are

$$\left.\begin{aligned} & r\frac{d}{dr}\left\{\frac{1}{r}\frac{d}{dr}\left[r^2\left\{\frac{Et}{1-\nu^2}\left[\frac{dv}{dr}-\frac{w}{R}+\frac{1}{2}\left(\frac{dw}{dr}\right)^2+\nu\left(\frac{v}{r}-\frac{w}{R}\right)\right]\right\}\right]\right\} \\ & \qquad + \frac{1}{2}Et\left(\frac{dw}{dr}\right)^2+Et\frac{r}{R}\frac{dw}{dr}=0 \\ & \frac{Et^3}{12(1-\nu^2)}\frac{d}{dr}\left[\frac{1}{r}\frac{d}{dr}\left(r\frac{dw}{dr}\right)\right]-\frac{Et}{1-\nu^2}\left[\frac{dv}{dr}-\frac{w}{R}+\frac{1}{2}\left(\frac{dw}{dr}\right)^2\right. \\ & \qquad \left.+\nu\left(\frac{v}{r}-\frac{w}{r}\right)\right]\left(\frac{r}{R}+\frac{dw}{dr}\right)-\frac{pr}{2}=0 \end{aligned}\right\} \quad (2.1$$

where the notation is the same as that in Figure 1.6. The relevant quantities as taken from Equations 2.1 are then given in Table 2.1.

Table 2.1. Relevant Physical Quantities for Elastic Buckling of Radially Loaded Spherical Cap

E = modulus of elasticity

ν = Poisson's ratio

p = applied pressure

R = shell radius

t = shell thickness

r = a measure of the horizontal extent, use r_0

The solution equation for critical pressure must be of the form

$$F(E, \nu, p, R, t, r_0) = 0 \tag{2.2}$$

If the thickness t and the modulus of elasticity E are taken as the dimensionally independent variables, then, according to Buckingham's theorem, Equation 2.2 can be reduced to

$$\Phi\left(\frac{p}{E}, \frac{t}{R}, \frac{t}{r_0}, \nu\right) = 0 \tag{2.3}$$

On the assumption that the critical pressure is in fact that of the first maximum on the load-deflection curve, Equation 2.3 can be solved for p/E in the form

$$p = E\phi\left(\frac{t}{R}, \frac{t}{r_0}, \nu\right) \tag{2.4}$$

An assumption of this nature is necessary because the criterion for shell buckling has not yet been agreed upon and the quantity p/E can be taken outside the functional expression only if the function is single-valued. Equation 2.4 could be written once for the prototype and once for the model:

$$\frac{p_{\text{prototype}} = E_p\phi\left(\frac{t_p}{R_p}, \frac{t_p}{r_{0p}}, \nu_p\right)}{p_{\text{model}} = E_m\phi\left(\frac{t_m}{R_m}, \frac{t_m}{r_{0m}}, \nu_m\right)} \tag{2.5}$$

Then if

$$\left.\begin{aligned} \left(\frac{t}{R}\right)_m &= \left(\frac{t}{R}\right)_p \\ \left(\frac{t}{r_0}\right)_m &= \left(\frac{t}{r_0}\right)_p \\ \nu_m &= \nu_p \end{aligned}\right\} \tag{2.6}$$

the observed model buckling pressure can be extrapolated to the prototype by

$$p_p = p_m \frac{E_p}{E_m} \tag{2.7}$$

If Equations 2.6 can be strictly satisfied and if the model loading does in fact simulate the prototype loading, then Equation 2.7 will give the correct value of the prototype buckling pressure. For all types of model problems the greatest difficulty encountered lies in the satisfaction of the particular equations like Equations 2.6. When these are not satisfied, real problems arise in the interpretation and extrapolation of the test data.

2.1.2 Model Material. Since the predominant state of stress in the thin shell is one of pure compression, it is not surprising that most of the shell roofs for buildings have been constructed of concrete. Steel reinforcing bars and prestressing wires are used to control shrinkage and temperature-induced tensile strains as well as to provide additional strength in certain boundary regions where the compressive membrane state is considerably altered by bending. Metallic shells are occasionally employed in civil engineering situations, but have a much greater application in the aircraft and missile industries.

In a structural model analysis which is intended to simulate a prototype over the complete loading range, up to and including material failure, the similitude restrictions which are placed on the model material are very severe. The stress-strain relations of the model and prototype materials must be similar throughout the entire strain range, up to and including failure. Considering only a uniaxial stress state such restrictions are as shown in Figure 2.3. If, on the other hand, one is interested only in prototype response at strain levels far below failure, then it may be possible to assume that the prototype behavior is elastic. The problem of instability in thin-shell roof structures may often be so approximated. For this elastic case, it is only necessary that the model also behave elastically throughout the pertinent strain range. Plastic materials, in addition to being elastic at low strain levels, possess three great advantages as potential model materials. First, they have a low modulus of elasticity, ranging from one or two million psi for some glass-reinforced materials down to a few thousand psi for some foamed plastics. The natural resins have a modulus of about 500,000 psi. Second, plastics are relatively inexpensive and easy to fabricate. The vacuum-forming technique for fabricating thin-shell models is particularly attractive. Finally, plastics have a high strength/stiffness ratio when compared to the common construction materials. Thus polyvinyl chloride plastic has a yield strength/modulus of elasticity ratio of approximately 0.02 in comparison to 0.0013 for concrete and steel. As a consequence of this fact, one can often extend the load-response function beyond that corresponding to the prototype and in this way reduce the errors due to "noise" in the experimental program. This technique is often used in obtaining load-deflection or load-strain curves. Of course, the extension feature is of no direct use in an elastic stability study, but there is an indirect benefit. Thus, if there is no material yielding even in the postbuckled condition, then the shell model can be used again and again. The experimental program of this monograph could not otherwise have been carried out.

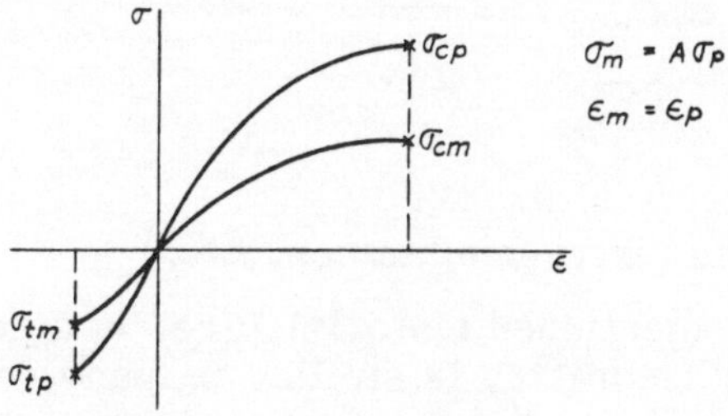

Figure 2.3. Necessary conditions for complete similarity of materials.

Plastics—General. The family of plastic materials can broadly be divided into two groups, thermoplastic and thermosetting. Thermoplastic resins are those which undergo no permanent change on heating. On continued heating above room temperature their tensile strength decreases and at temperatures of 150 to 300° F they become quite rubbery. At even higher temperatures they melt. This situation is shown qualitatively in Figure 2.4. At the elevated temperatures they can easily be formed into a variety of shapes which they retain on cooling. Of course, the process can be repeated and the plastic can be remolded into some new shape. By distinction, a thermosetting resin is one which does not possess this property of being able to be reformed

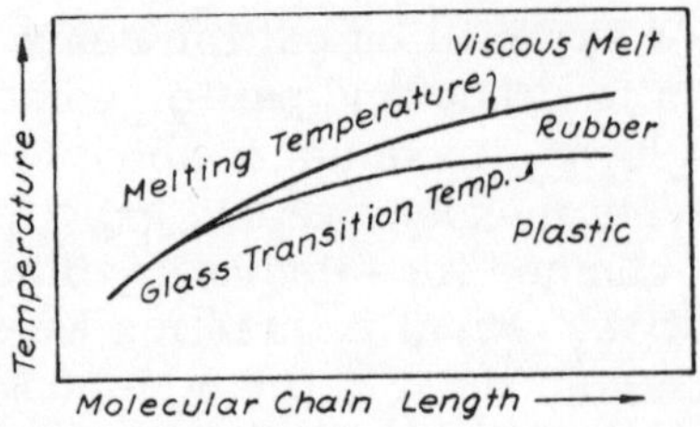

Figure 2.4. Temperature sensitivity of linear polymer materials.

at will under elevated temperatures. Once the thermoset has achieved its rigid form it maintains that shape. A very general grouping of the common plastic materials might be as follows:

<u>Thermoplastics</u>

Polyethylene
Polystyrene
Cellulosics (cellulose nitrate, cellulose acetate, etc.)
Vinyls (polyvinyl chloride, polyvinyl acetate, etc.)
Acrylics (methyl methacrylate)

<u>Thermosets</u>

Phenolics
Epoxies
Polyesters
Silicones
Amino plastics (urea, melamines, etc.)

"All of the aforementioned materials are, in a sense, organic chemicals. Their chemistry is similar in many respects to that of other organic chemicals such as sugar, dyes, or aspirin. The unique properties of these resins, and the properties which they have in common, result from the fact that their component molecules are tremendously large. These very large molecules are, however, made up of relatively simple repeating units. If the polymer contains perhaps 500 or more repeating units, it becomes known as a "high" polymer. It is these polymers which in general comprise the resin molecules.

"Molecules of the thermoplastic resins are characteristically high polymers with long, continuous carbon-atom chains for a molecular framework. The chemical formula for the resin becomes the formula for the monomer, but multiplied by some number representing a degree of polymerization. Essential differences between thermoplastic resins are associated with the specific chemical groupings attached to the carbon-atom chain. These groupings may vary considerably, and this variation permits the production of tailored molecules designed to give resins for specific purposes.

"Molecules of the thermosetting resins are usually quite similar to those of the thermoplastic resins before molding. But the setting process is accomplished by binding together chemically adjacent molecular chains into a complex three-dimensional network. The molecules thus become even larger and still more complex. Indeed, it has been observed that all of the atoms in an individual thermoset specimen may be bonded together chemically, and a specimen may thus conform to the classic definition of a single molecule."[23]

When there are so many plastic materials it is not at all sufficient to say that one is going to use plastic as a model material. The choice of a specific type must be based upon a rational consideration of the engineering properties of the various types, because these engineering properties do vary significantly from one plastic to another.

To speak of the tensile strength or the modulus of elasticity of a certain plastic material without first specifying the applicable conditions of temperature, relative humidity, rate of loading, etc., is to speak without meaning. The mechanical properties of plastics may be significantly affected by the aforementioned factors. However, for any particular set of conditions it is true that tests run on various samples of a commercially produced plastic such as Plexiglas will yield consistent results. In Figure 2.5 the uniaxial stress-strain diagrams are shown for several

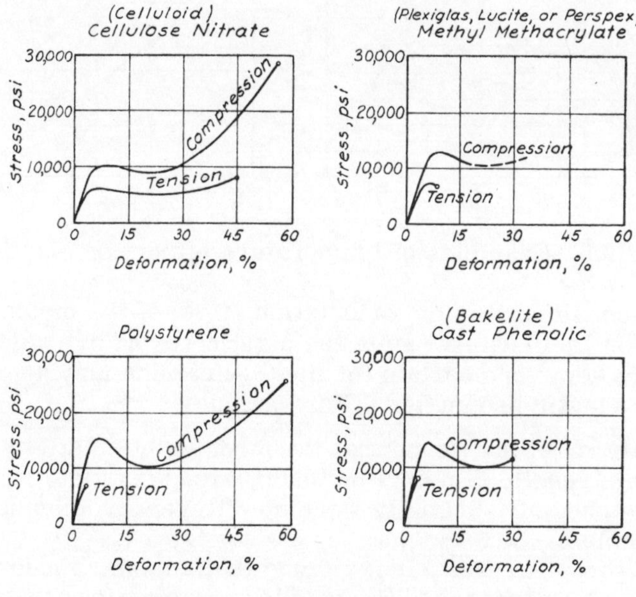

Figure 2.5. Stress-strain diagrams for some plastic materials at room temperature.[27]

plastics. The deformations involved may rather arbitrarily be divided into three types. First, there is a straightening of valence bond angles between the atoms of any molecule. This type of deformation is nearly instantaneous and is linearly elastic. The second type of deformation is associated with coiling or uncoiling of the molecular chains. There is no permanent change in intermolecular arrangement, and consequently, it would seem that these deformations are all recoverable although perhaps not instantaneously. This type of deformation is particularly prevalent in thermoplastics that have been heated above their glass-transition temperature. Thus the vacuum-drawing operation and the so-called "memory effect," whereby a vacuum-formed part returns to its original shape upon reheating, depend upon such uncoiling and coiling, respectively. The third and last type of deformation actually involves intermolecular slippage. Such deformation cannot be recoverable.

Temperature and relative humidity may have a significant effect upon the mechanical properties of plastic materials. Figures 2.6 and 2.7 show typical effects. Strain rate also influences the stress-strain curves, and although data are lacking in this respect, Figure 2.8 indicates the type of behavior to be expected. As might be expected, increasing the strain rate has the same effect as lowering the temperature. Yet another factor which

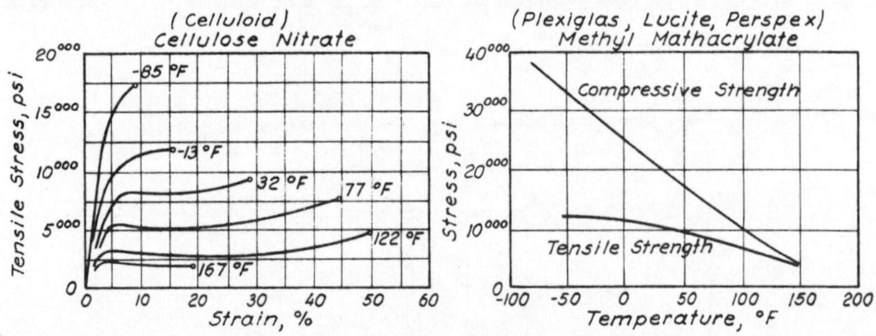

Figure 2.6. Examples of temperature effects on plastics.[27]

must be considered is the orientation effect of the molecular chains. In their natural state the plastic resins are relatively isotropic; however, certain forming operations may appreciably alter the elastic properties. For example:

1. Axilrod, et al.[1] found that the modulus of elasticity and tensile strength of methyl methacrylate (Plexiglas) vacuum-stretched 50% biaxially were no different from unstretched samples.
2. Northrop Aircraft, Inc.[32] found that uniaxial vacuum stretching increased the tensile strength along the flow lines and decreased the strength across the flow lines for methyl methacrylate.

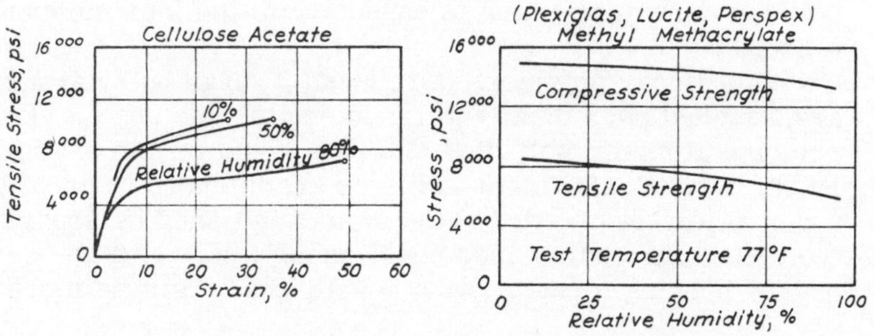

Figure 2.7. Examples of relative humidity effects on plastics.[27]

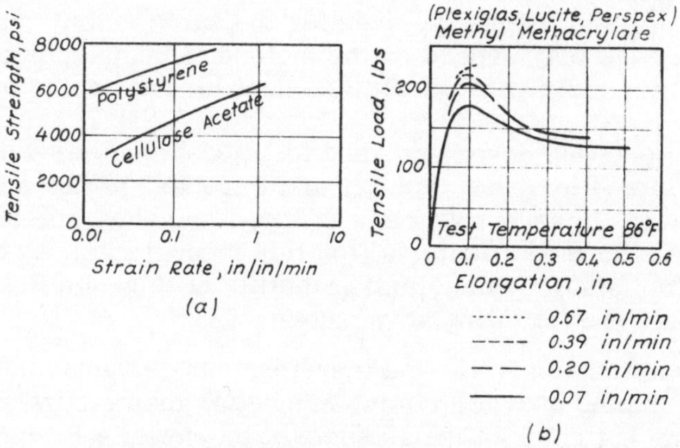

Figure 2.8. Representative effects of strain rate on tensile strength.[23, 28]

3. Bailey,[2] with polystyrene, had the same experience as Northrop but, unlike Axilrod, found that the strength of biaxially stretched material increased.

Of vital concern, whenever a plastic is to be used as a model material, is a consideration of the creep characteristics of the plastic. By comparison with other common construction materials, plastics are very creep-sensitive. The mechanism of creep in plastic materials is not well understood,[13] but it is generally agreed that creep involves a rupture or slippage of certain secondary bonding forces, this occurring by a sliding of one molecule past another or by an uncoiling (coiling) of the molecular chains. Generally:

1. Thermoplastics are subject to greater creep than thermosets. This is so because the thermoplastics have a linear

polymer structure which is conducive to the intermolecular slippage.

2. Creep of linear polymers is reduced if large bulky atoms are attached periodically along the molecular chain. The chlorine atoms in polyvinyl chloride would be one example.
3. Plasticizer (flexibilizer) additives are intended to provide a more flexible plastic. This is accomplished by separating the molecular chains and, consequently, lowering the secondary bonding forces. As a result, the plastic is more creep-sensitive.

Based only on what has been presented here, the stress-strain-time relations in plastic materials are quite complex. Furthermore, it may be said that varying the "mix" of any plastic material will produce additional changes. That is, amounts of plasticizers, filler materials, etc., affect properties as well as the basic resin itself. Finally, even for the same "mix," mechanical properties may depend on the method of manufacture. The experimenter must very carefully select his particular plastic material.

Various plastics have been used for model analysis with methyl methacrylate (Plexiglas, Lucite, and Perspex) perhaps being used the most in recent years. A Type I polyvinyl chloride was selected as the model material for this monograph. As a model material for studying the elastic stability of thin-shell structures it possesses the following advantages:

1. It is a thermoplastic and therefore can be vacuum-formed. Its forming characteristics are better than methyl methacrylate but not as good as some of the newer ABS (acrylonitrile-butadiene-styrene) plastics.
2. Its stress-strain characteristics are quite linear up to the 2000 psi level. Further, the quality control of the manufactured sheets is very good and the vacuum-forming operation does not induce large changes in the modulus of elasticity.
3. The thickness variation in any one sheet is ± 5% maximum which is considerably better than in Plexiglas. The reason for the better thickness control is that PVC sheets are calendered while the Plexiglas ones are cast between two plates.
4. At room temperature and 1000 to 2000 psi stress levels, creep strains in one hour are on the order of 5% of the initial strain. In the thermoplastics this is quite low. Plexiglas will creep almost twice as much as rigid PVC.
5. Its cost is modest—certainly not the controlling cost factor in a model study.
6. Sheet thicknesses are available down to 0.010 inch, although

for Boltaron 6200, the specific material used in this study, minimum sheet thickness is 0.030 inch. Plexiglas is not available in such a range of thicknesses.

The reasons for selecting PVC over Plexiglas are noted in points 1, 3, 4, and 6. Of course, all thermosetting plastics were ruled out because it was desired to vacuum-form the model shells. Polyethylene and the cellulosics must be ruled out because of their creep characteristics, the polyamides (nylon) are not suitable for vacuum-forming even though they are thermoplastics and the new acetal resins (Dupont's Delrin) are not available in sheet form. Still it is not possible to state without qualification that Type I PVC is the best choice. A medium impact ABS resin (Boltaron 6500) would have better forming characteristics, a lower modulus of elasticity (320,000 psi), but more creep. Lexan, a polycarbonate resin developed by General Electric in 1958, is reported to have a lower modulus of elasticity (375,000 psi) and excellent creep characteristics, although the cost is much higher than PVC. What can be said is that Type I PVC does not have any serious disadvantages when compared to the other plastics. Perhaps, with another material, a gain could be had in achieving slightly improved molding characteristics, a slightly lower modulus and/or slightly improved creep properties, but that is all.

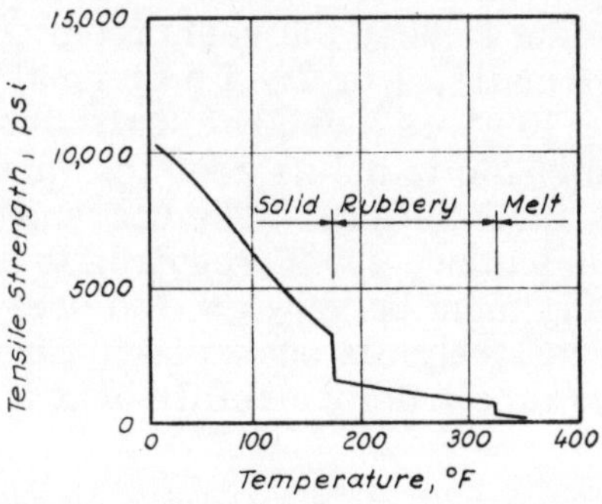

Figure 2.9. Behavior of unplasticized polyvinyl chloride.[23]

Vinyl chloride monomer has the chemical formula $CH_2 = CHCl$. The pure polymerized resin which is known as unplasticized polyvinyl chloride is clear and rigid. Figure 2.9 illustrates the effect of temperature on the tensile strength of unplasticized polyvinyl chloride. It would appear that a relatively broad temperature range exists over which vacuum-forming would be successful; however, the fact is that only a narrow range exists wherein the material elongates sufficiently to make possible the forming of a broad range of parts. Additionally, unplasticized polyvinyl chloride has a low impact strength. To improve forming ease and to increase impact strength, it is customary to incorporate in the polyvinyl chloride resin as little as 5 to 15% of a "rubber-like" material. Common additive materials have been vinyl acetate $CH_2 = CH - C_2H_3O_2$; acrylonitrile, $CH_2 = CHCN$ and vinyl stearate, $CH_2 = CH(C_{12}H_{23}O_2)$. The resulting mixtures are ordinarily available in two grades, a Type I normal impact grade and a Type II high impact grade. The mix of these grades would ordinarily be about 95 PVC to 5 and 85 PVC to 15 respectively, and it is common for each of them to be referred to as rigid polyvinyl chloride.

The particular Type I material used in this investigation is manufactured by Bolta Products, Division of General Tire and Rubber Co.[5] It is sold under the trade name of Boltaron 6200 and incorporates acrylonitrile-butadiene-styrene (a synthetic rubber known as an ABS plastic) as the copolymer. Table 2.2 compares some of the properties of Boltaron 6200 with other formulations made by Bolta Products. Thus the "mix" for Boltaron 6200 consists of pure unplasticized PVC and ABS resin in the ratio of 95 to 5. As stated previously the unplasticized PVC is clear whereas the addition of the ABS resin leads to a yellow translucent plastic. Actually carbon black and titanium oxide are also included as pigments, and the finished product is actually opaque and gray in color. About 1% calcium stearate is included as a lubricant to keep the resin from sticking to the calendering rolls. Finally, 1 or 2% of a thermal stabilizer is included. Figure 2.10 gives a diagrammatic insight into the reasons why Boltaron 6200 is not as good for vacuum-forming as some other Bolta formulations. The decrease in ultimate elongation which occurs above 230 °F requires that temperature control during forming must be very good to prevent tearing in places where large stretch is occurring. The 6100 and 6500 formulations maintain their deformability over a large temperature range.

Table 2.2. Comparisons of Different Boltaron Formulations[5]

Boltaron	PVC - ABS %	Tensile Modulus of Elasticity (psi)	Tensile Strength (psi)	Ranked Creep-Sensitivity	Ranked Vacuum-forming Ease	Ranked Cost
6200	95-5	420,000	7500	Best	Worst	3rd Lowest
7200	85-15	250,000	6000	2nd Best	3rd Best	Highest
6500	50-50	260,000	5800	3rd Best	2nd Best	2nd Lowest
6100	0-100	200,000	3500	Worst	Best	Lowest

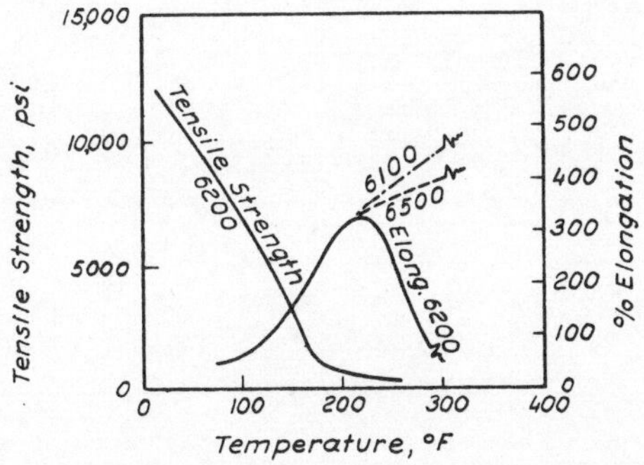

Figure 2.10. Vacuum-forming characteristics of Boltaron plastics.[5]

In Figure 2.11 a uniaxial stress-strain curve in tension and compression is given for Boltaron 6200. It should be pointed out that this curve should not be relied upon for any and all Boltaron 6200 applications. Tensile strengths range from 13,000 to 7000 psi depending upon technique of manufacture. Calendered sheets, for example, will run around 9000 to 10,000 psi as shown in Figure 2.11. Figure 2.12 shows an untested sample in comparison to a severely necked down specimen. Figure 2.13 gives some other properties for rigid polyvinyl chloride. The graph for modulus of elasticity versus temperature shows a slope of about 1000 psi per °F in the room temperature range. No quantitative data could be found on the effects of relative humidity but it is not as sensitive as Plexiglas, the characteristics of which are shown in Figure 2.7.

In order to evaluate Boltaron 6200 PVC as a model material for studying the elastic stability of prototype structures, one must consider the stress-strain relations of the prototype material. If the prototype material is concrete, the prototype material stress-strain curves might be as shown in Figure 2.14. If the strain levels at failure are small enough, plastic will satisfactorily simulate concrete. Of the twenty 18-inch radius shells tested under air pressure and reported in Chapter 4, shell 5-1 underwent the largest strains prior to buckling. If this model result were to be extrapolated to a prototype, the similarity in material properties would be as shown in Figure 2.15 if the 5000 psi concrete from Figure 2.14 is imagined as the prototype material. When it is considered that the twenty model shells buckled at pressures of from 0.50 to 1.07 psi one could extrapolate these results to an imaginary concrete prototype by

$$p_{prototype} = p_{model}\frac{E_p}{E_m} = 144(0.50 \text{ to } 1.07)\frac{3,500,000}{450,000}$$

$$= 560 \text{ to } 1200 \text{ psf}$$

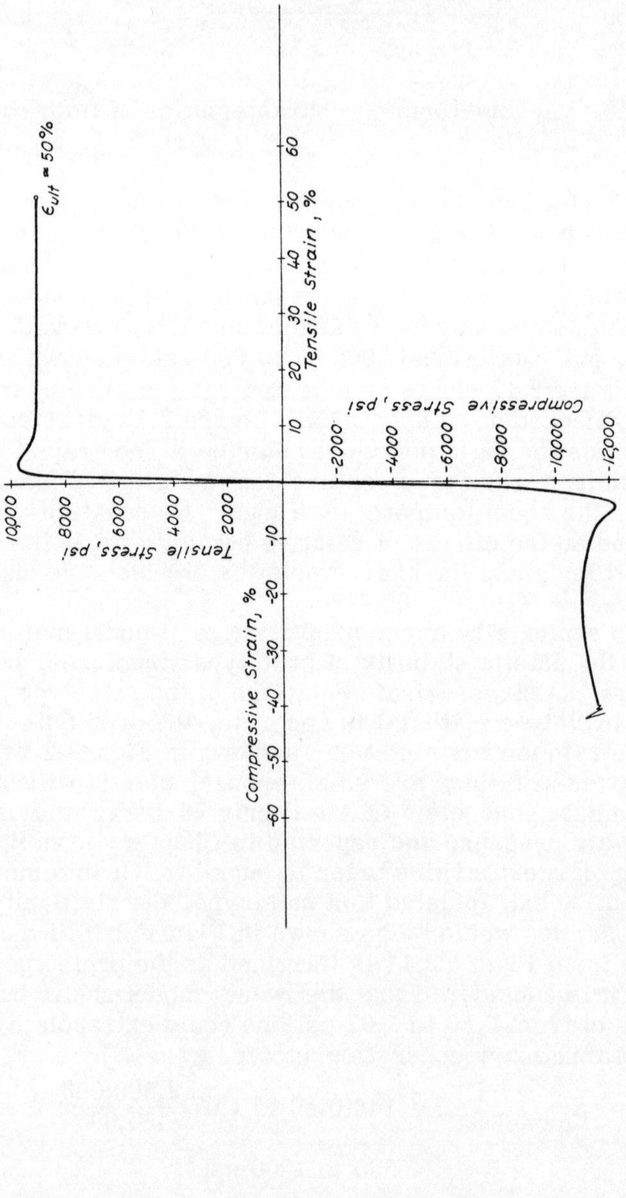

Figure 2.11a. Uniaxial stress-strain curve for Boltaron 6200 PVC.

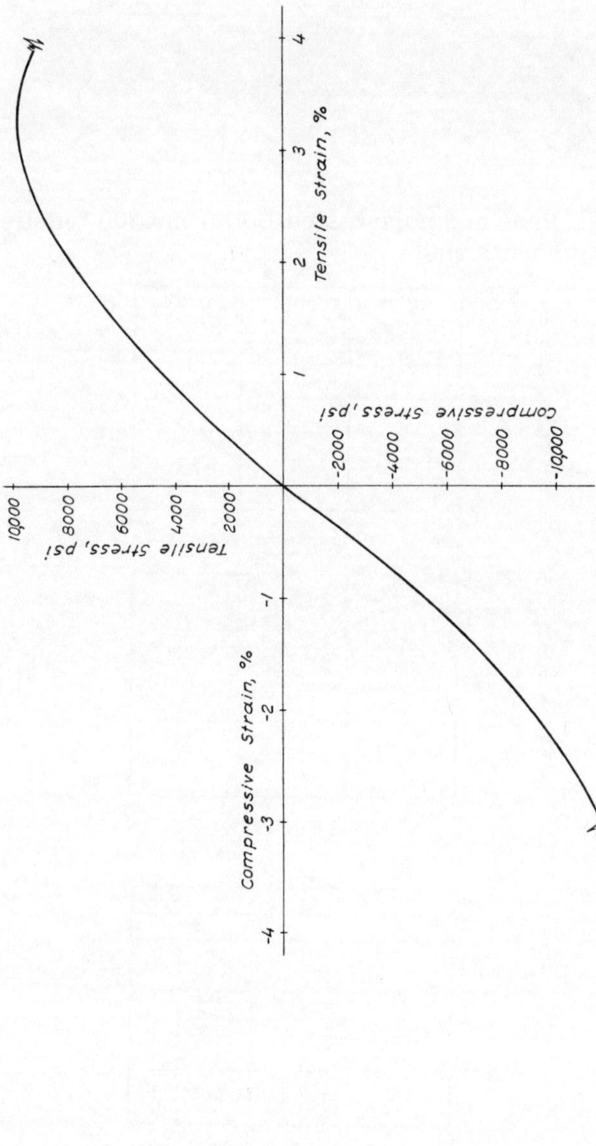

Figure 2.11b. Uniaxial stress-strain curve for Boltaron 6200 PVC.

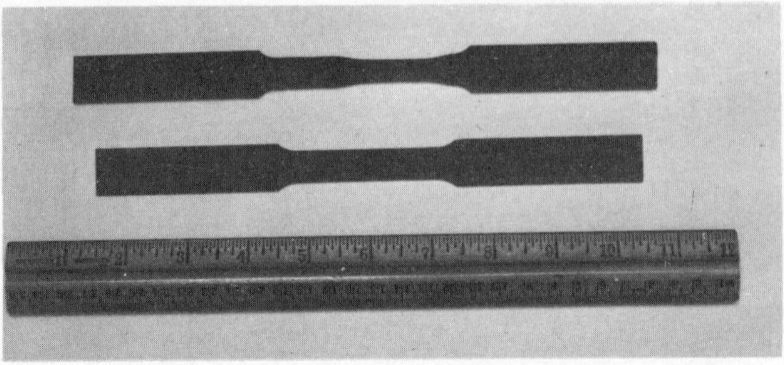

Figure 2.12. Pre- and post-tested Boltaron 6200 tensile specimens.

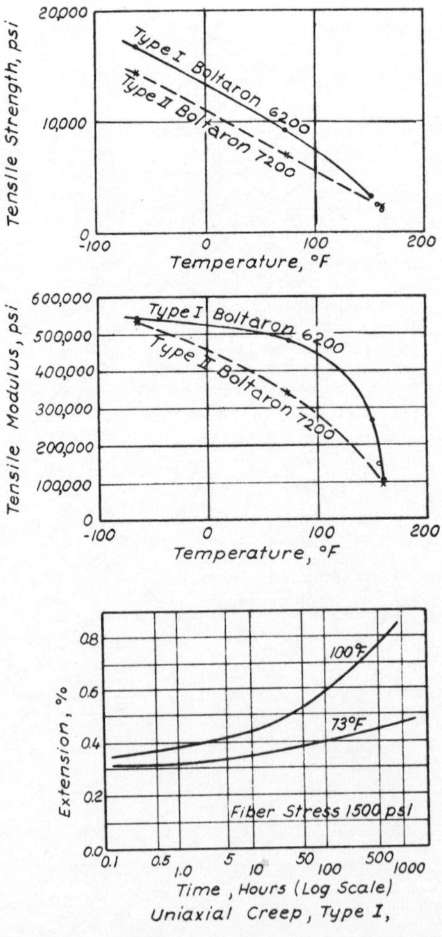

Figure 2.13a. Some properties of Type I PVC.[5]

		Boltaron 6200 PVC
General	Nature	Thermoplastic
	Color	Unplasticized PVC is Clear 6200 is Opaque and Gray
	Specific Gravity	1.43
	Available Shapes	Calendered Sheets ($\frac{1}{32}$, $\frac{1}{16}$, $\frac{3}{32}$, $\frac{1}{8}$, $\frac{1}{4}$, $\frac{1}{2}$"), Rods, Tubes
	Thickness Variation in Preformed Sheets	± 10 % Max.
	Cost	$\frac{1}{32}$" - $0.55 /ft^2 $\frac{1}{16}$" - 0.90 $\frac{3}{32}$" - 1.20 $\frac{1}{8}$" - 1.50
	Jointing Characteristics	Can be Welded Using PVC Rods and Heated Inert Gas
	Shrinkage	1% @ 270 °F
	Water Absorption (ASTM D570)	0.15 % in 24 Hours
Mechanical	Order of Magnitude Tensile Properties (ASTM D638) 73 °F	9500 psi @ Upper Yield Strain @ Upper Yield = 3.2 % Ultimate Elongation ≈ 50 %
	Order of Magnitude Compression Properties (ASTM D695) 73 °F	12000 psi @ Upper Yield Strain @ Upper Yield = 5 %
	Uniaxial Creep	0.000400 "/in -1500 psi @73 °F, 24 Hours
	Modulus of Elasticity (ASTM D638, D695, D790)	460,000 ± 30,000 psi @ 73 ° F
	Poisson's Ratio	~0.38
Thermal	Vacuum-Forming Temperature	185 - 260 °F
	Thermal Conductivity	$1.16 \frac{BTU\ (IN)}{(Hour)(Ft^2)(°F)}$
	Coefficient of Expansion	3.7×10^{-5} in/in/°F @ 70 °F
Optical	Refractive Index	1.55 for Unplasticized PVC
	Luminous Transmittance	Dependent on Composition 6200 is Opaque

Figure 2.13b. Some properties of Boltaron 6200 PVC.[5]

Of course, one might wish to use a long time modulus of elasticity for the prototype, but still it becomes apparent that buckling of thin-shell roof structures is not usually a problem unless the shell is quite flat or very large and thin.

2.2 Fabrication

There are several ways in which thin-shell plastic models could be made. In a model analysis the experimenter is usually interested in fabricating only one or at most a few models. This fact is responsible for his rejection of what in one sense are the most powerful molding techniques available. Commercially, millions of pounds of plastic resins are compression, transfer, and injection molded into every conceivable shape and, within limits, size. These three general techniques employ very high pressures and temperatures in conjunction with very elaborate molds (pressures reach 4000 pounds per square inch of mold surface in compression molds and 25,000 pounds per square inch of plunger area in injection molds, temperatures reach 600 °F, and molds may cost $10,000 apiece). Clearly then, these commercial

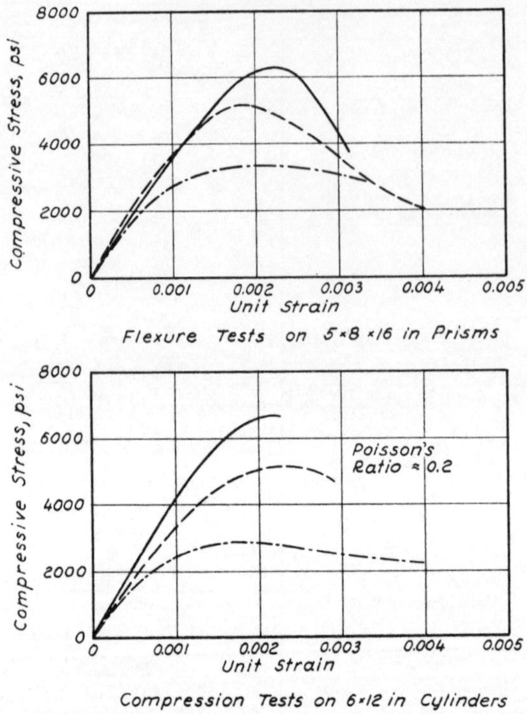

Figure 2.14. Standard concrete compression stress-strain curves at age 28 days.[18]

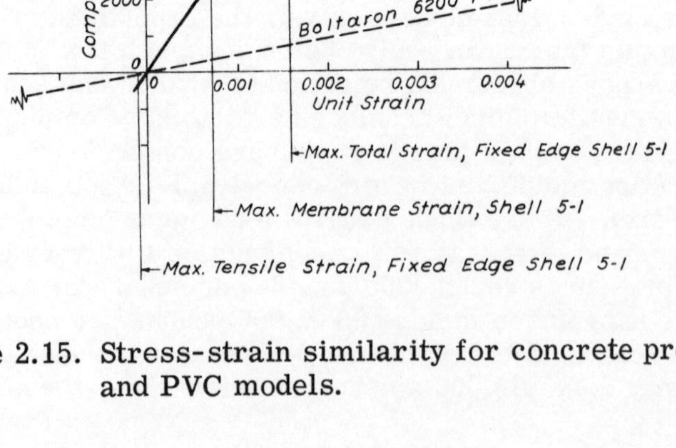

Figure 2.15. Stress-strain similarity for concrete prototype and PVC models.

operations are not suited to the fabrication of a single or even a few items. For shells of constant thickness the process of vacuum-forming rigid thermoplastic sheet materials is feasible and economical. This technique, which consists of heating the plastic sheet above its glass transition temperature and then pressing this sheet against a prepared mold, was used for the fabrication of every model shell. The most important step in the fabrication process is the making of an accurate mold. Metal, wood, plaster, certain plastics, etc., are suitable mold materials; however, only hardwood and gypsum plaster were considered.

A natural question that might be asked is whether the thickness changes that must be a consequence of the vacuum-pressing process are tolerable. This is a two-sided question. First, what sort of thickness variations result from such pressing and second, how do shell thickness variations affect the structural behavior that is to be investigated? With regard to the first part, preliminary experimentation showed that models of the selected spherical cap could be manufactured such that less than ± 10% maximum thickness variations could be achieved. This was considered satisfactory. The experiments themselves are to throw some light on the second part of the question.

After the shells would come from the vacuum-forming machine, only trimming would be required to give the finished model.

2.3 Loading

The loading system for a structural model incorporates the means of boundary support as well as the actual provision for load application. In studies involving an elastic instability failure there are several factors which must be considered. Of critical importance is the fact that the loading system must not restrain the model. In addition the load must be applied simultaneously, this being true whether or not the load is symmetrical or unsymmetrical. Finally, since the behavior is to be elastic and models are to be recovered for further testing, some provisions must be made to "catch" the model before it is destroyed.

One of the heretofore unmentioned reasons for the choice of the spherical cap as a vehicle to study the model reliability was that its continuously supported edge makes air pressure loading particularly easy. The pressure loading will give a uniform loading, will not restrain the model, and will compare with previous theoretical and experimental results.

Of course, radial pressure does not simulate a gravity loading unless the shell is very flat. On steep slopes it would be necessary to provide vertical load, and this is most easily accomplished by hanging weights. When weights are used to apply load, several potential difficulties arise, namely: (1) a discrete loading is substituted for a continuous loading, (2) the loading system may restrain the model, and (3) provision must be made for applying the loads simultaneously. To evaluate the effect of these three factors, a series of weight-loading tests on shells previously loaded by air pressure were planned.

The model shell geometry was selected in order that the critical buckling pressures would be of the order of 1/2 to 1 psi. Air pressures of 1/2 to 1 psi can be accurately measured using a water manometer. An adequate vacuum system was available and the apparatus depicted in Figure 2.16 was planned. For the weight-loading tests holes were drilled through the model on a 1-inch surface grid and weights were hung from each string. On the 18-inch radius domes this meant 241 load points. One-half inch diameter steel weights were available in nominal sizes of 1, 1/2, 1/4, and 1/8 pounds. For smaller increments, tire chain monkey links weighing approximately 0.026 pound were used. The handling of 200 pounds in weight loading is tedious and time-consuming but entirely feasible.

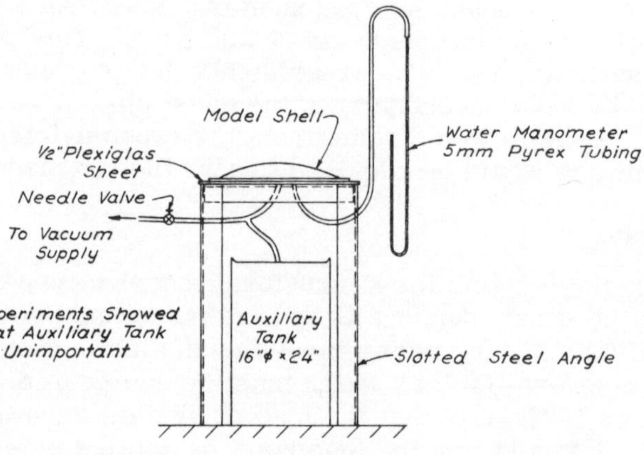

Figure 2.16. Schematic diagram of vacuum loading system.

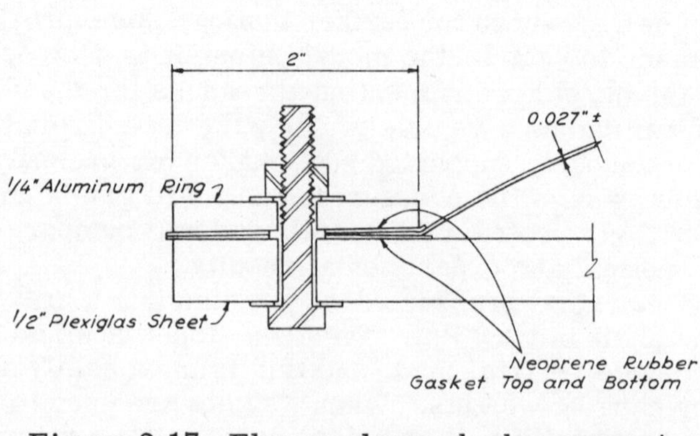

Figure 2.17. Flange-clamped edge support.

For the boundary supports, the initial thought was to vacuum-form the domes with a planar flange around the edge. This edge could then be clamped as shown in Figure 2.17. The test results with clamped flanges indicated the deleterious effect of flange clamping; consequently, the flanges were removed and the tests repeated with the edges encased in a bed of epoxy cement.

2.4 Instrumentation

Instrumentation was not one of the controlling considerations in the experimental program. Three things were to be determined: (1) measurement of the critical pressure, (2) visual observation of the buckle position, and (3) certain attendant geometric measurements such as shell thickness and shell geometry. A water manometer can accurately measure the air pressures and for the weight loading one merely counts the weights. Shell thicknesses were measured with the Ames dial gage apparatus shown in Figure 2.18. Edge thicknesses were also measured with a micrometer and they checked the Ames dial readings. It is difficult to measure the geometric shape of the model shells. Although it was not used, a photogrammetric technique would be an accurate procedure and more importantly would establish the entire surface. The technique used in this monograph involved placing an accurately machined template over the shell until it rode on the high spots. The gap between the template and the shell was then measured to the nearest 0.001 inch. This was done along two great circles oriented at 90° to each other.

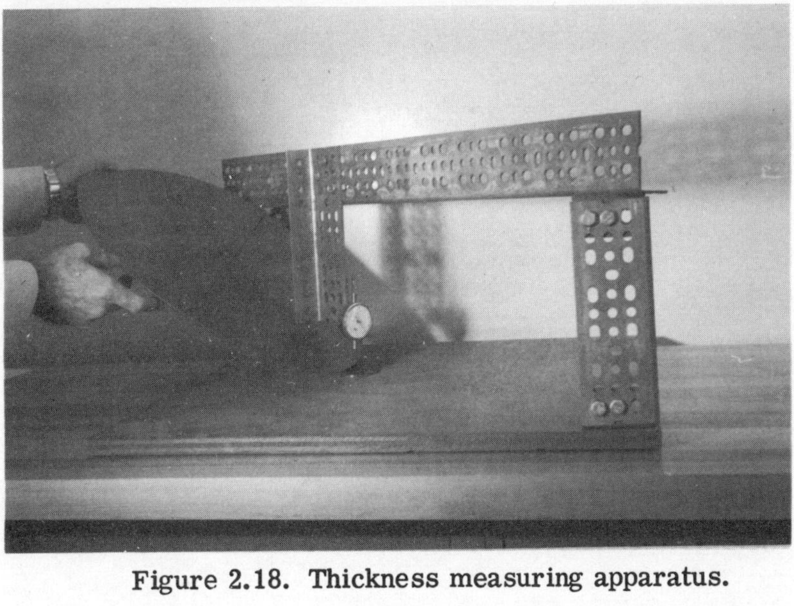

Figure 2.18. Thickness measuring apparatus.

2.5 Interpretation

The experimental studies contained herein are not intended to yield information which can be extrapolated to any particular prototype. Rather, this is a research study to investigate some factors which could influence the reliability of any single model test. The interpretation will then be to evaluate the data in light of the questions asked in the second paragraph of Section 2.1.

Chapter 3

ERROR ANALYSIS

The purposes of this chapter are to point out that any structural design problem is really probabilistic in nature; to indicate how certain elementary principles of statistics and probability can be applied in the experimental design process; and to conclude that, though the statistical theory is a very powerful tool, the major effort in an experimental design study should be directed toward eliminating systematic errors from the experimental result. The emphasis which is placed upon systematic error elimination arises because the theories of probability and statistics have been developed for random phenomena, and experience shows that many model studies are dominated by systematic errors rather than random ones.

3.1 Structural Design as a Problem in Probability

The basic problem of the structural design engineer is to combine structural forms and materials in such a way that the resulting structure will "safely" resist the environmental influences to which it is subjected. Although the concept of safety lies at the very heart of the design process, it would seem that structural engineers have not properly taken account of the fact that the resultant response of a structure when subjected to various environmental conditions is really a probabilistic problem incorporating the random characteristics inherent in loading, material strength, structure geometry, etc. Thus, if two supposedly identical structures were constructed, the engineer would not expect their response characteristics to be identical but rather would expect to notice some variation in their response. Suppose, for example, that one were concerned with an ordinary reinforced concrete cantilever retaining wall which was to be placed at various locations along a highway. The engineer prepares a single design calling for certain heel and toe lengths, stem thickness, stem reinforcing steel, concrete strength, etc., and the wall is constructed at seven locations along the highway.

Table 3.1 describes in a hypothetical way some final conditions at each location. Several things can be noted with regard to this table. The stem thickness is not constant, there being relatively small deviations around what seems to be an average value of about 12 1/8 inches. Likewise there is variation in concrete strength about an average value of 3300 psi. The overturning force that the wall must resist is a function of the properties of the backfill materials, method of placement, provision for drainage, etc., and one should expect from the table that the magnitude and distribution of the pressures which are exerted on the seven walls are perhaps considerably different. The support or boundary conditions are a function of the foundation materials which again differ rather widely.

Table 3.1. Hypothetical Prototype Retaining Wall Conditions

Wall	Stem Thickness (in.)	Concrete Strength (psi)	Type of Backfill	Foundation Material	Distance of Main Stem Reinforcing Steel from Rear Wall Face (in.)
1	12	3300	Granular	Dense silty sand	3
2	12 3/8	2900	Compacted silty sand	Dense silty sand	3
3	12 1/8	3100	Compacted silty sand: no drainage	Dense silty sand	9
4	11 3/4	2900	Dense sand	Loose sand	3
5	11 7/8	3600	Sandy silt	Dense sandy silt	3 1/4
6	12 1/2	4100	Dense sand	Clayey sand	3 1/2
7	12 1/4	3200	Granular	Miscellaneous fill	3

Suppose that a short distance further along this highway an overpass bridge was to be constructed, the abutments of which were to be faced with an expensive stone facing. A retaining wall is to join with this abutment and the architect is quite concerned with regard to the details of the joint. As the engineer, you are asked to estimate the outward deflection of the retaining wall at this joint. The customary procedure would be to assume values for the series of pertinent physical quantities, that is, stem thickness, concrete modulus of elasticity, steel modulus of elasticity, magnitude and distribution of applied pressures, etc. Then a computation would be made using a postulated mathematical model and a certain specific deflection would be determined—say 3/4 inch. The engineer might then predict that the deflection would be between 1/2 and 1 inch, where the dispersion about his computed value of 3/4 inch was rather arbitrarily estimated in such a way as to satisfy his fears that some of his assumptions may have been in error. The fact that the engineer feels the need to state a range implies that he realizes the probabilistic nature of the problem even if he does not know how to deal with this nature formally. Alternatively, suppose that the deflection of the walls of Table 3.1 have been measured. One would surely not expect that all seven walls would deflect the same amount. In fact wall 3 would likely fall flat on its face on account of the blunder of the construction workers who for some strange reason placed the main stem reinforcing steel in the front face. The other six walls might deflect in varying amounts depending upon the factors listed in Table 3.1 and many other factors not listed. The engineer might attempt in some way to relate the wall at the abutment to the six remaining measured walls and in this way predict a deflection based upon the six previously observed values. If a logical quantitative answer can be given, then this answer is directly tied to the random and systematic deviations associated with the observations recorded on walls 1, 2, 4, 5, 6, 7, and hence the quantitative answer must necessarily be of the form that with a 95 or 90 or 60% chance of being correct the deflection will be between 0.6 and 0.8 inch.

To close this introductory discussion, it should be noted that the model engineer is faced with additional difficulties. Suppose that he has been asked to determine, by small scale model studies, an estimate of how much prototype wall number 8 will deflect. He must first decide what he is going to try to reproduce in the model. Having done this, he proceeds to make a number of model studies and arrives at a table similar to Table 3.1, only this time the entries are for model walls not prototype walls. He is now faced not with the problem of estimating the deflection of another as yet unconstructed model wall, but rather with the problem of estimating the deflection of the prototype wall.

Two points should be made in this regard. First, it has been mentioned in Chapter 2 that one of the prime advantages of the model or experimental method of design lies in the fact that one does not need to know the complete analytical formulation of the problem in order to proceed with a meaningful model design. For the general model problem it will be shown that the behavior of the prototype cannot be predicted on a probabilistic basis. On the other hand, it is still of great interest to be able to deduce a "best" estimate of the prototype outcome, and in this respect careful model studies can be extremely valuable. Second—and with the previous point in mind—the model counterpart to Table 3.1 again will be subject to blunder, random and systematic errors. For design purposes, a result within 10 or 15% would be quite acceptable. It is particularly important that every effort be made to eliminate blunder and systematic errors from the model study, for it is these types of error which may affect results by a factor of 100 or 200%.

3.2 Errors in Structural Model Investigations

In Chapter 2 it was noted that the structural model analysis can be considered as incorporating a sequence of five steps, namely: (1) planning, (2) fabrication, (3) loading, (4) recording the data, and (5) interpretation and/or extrapolation to the prototype. Errors may enter in each of these five steps and the following list of possible error sources is given not with the intention of being exhaustive but merely illustrative.

Planning

1. Mistake in dimensional analysis
2. Failure to recognize a relevant variable

Fabrication

1. Geometry: thickness, length, etc.
2. Material properties
   a. Poisson's ratio, for example, $\nu_{plastics} = 0.3\text{-}0.5$ whereas $\nu_{concrete} = 0.2$
   b. Modulus of elasticity
   c. Complete stress-strain-time characteristics
   d. Coefficient of thermal expansion
   e. Density
   f. Microscopic and macroscopic structure
   g. Creep characteristics
   h. Initial stresses

Loading

1. Boundary conditions
2. Magnitude of load
3. Direction of load
4. Distribution of load
5. Time history of load
6. Errors associated with hanging weights at discrete points to make up gravity load deficiency

Recording

1. Electric resistance gages
   a. Incomplete bonding of adhesives
   b. Chemical attack on plastics by bonding adhesives
   c. Temperature compensation
   d. Calibration errors
   e. Inherent recording instrument error
   f. Gage factor error
   g. Transverse sensitivity
   h. Current heating effect on plastic materials
   i. Gage stiffening of plastic materials
2. Displacements
   a. Judgment errors in smallest division of instrument
   b. Support system of recording device not compatible

with magnitude of displacements
   c. Ditto circuitry, calibration, etc. errors listed under electric resistance gages
3. Pressure
   a. Meniscus corrections in a liquid manometer

Interpretation
1. Incorrect assumption for transformation of measured surface strains into stresses; e.g., assumption of plane strain instead of plane stress.
2. Slide rule error in reduction of data

This listing includes a wide variety of errors, some integration of which determines what is commonly referred to as experimental error. In a more specific sense, however, each of the errors listed above may be considered to fall into one of three general error categories: (1) blunders, (2) random errors, and (3) systematic errors.

3.2.1 Blunders. This type of error has no place in a scientific experiment. They are outright mistakes and should be eliminated by care and repetition of measurements. Examples of blunders would be:

1. Incorrect logic in dimensional analysis
2. Misreading an instrument
3. Mistake in dimensional units
4. Mounting a strain gage in incorrect position

3.2.2 Random Errors. It is impossible to give a rigorous operational definition of random; however, the nature of the concept is associated with the fact that a random phenomenon is an empirical phenomenon characterized by the property that its observation under a given set of circumstances does not always lead to the same observed outcome but rather to different outcomes in such a way that there is "statistical" regularity between these different outcomes. In view of this vagueness, it is not surprising that several meanings have been advanced for random errors. The differences in such meanings are rather subtle, however, and one can think of a random error as the difference between a single measured value and the "best" value of a set of measurements whose variation is random. What constitutes the "best" value depends on one's purpose, but here the best value will always be taken as the arithmetic mean of all the actual trial measurements. It should be noted that the algebraic sign of a random error can be either positive or negative.

Random errors may arise in two rather different contexts. First, there are random phenomena associated with the statistical nature of the physical model or the property being measured.

For example, the depth of 1000 18 WF 50 steel beams would not each be expected to equal the nominal value of 18.00 inches. In fact the steel companies specify a tolerance of $\pm 1/8$ inch so that one might expect to find a range of depths, perhaps the great majority lying between 17.9 and 18.1 inches but with an exceptional one falling outside these limits. Similarly, the yield stress in a certain portion of each of the 1000 beams would vary over a range of values, perhaps between 28,000 and 48,000 psi. Second, random errors may be introduced directly as a part of the measuring process. Examples of these errors would be the variation inherent in estimating the smallest division on some measuring instrument or the fluctuation in apparent strain due to random supply voltage changes in an electric resistance gage circuit.

3.2.3 Systematic Errors. Suppose now that the "best" value of the depth of the 18 WF 50 beams is 17.99 inches. Now someone comes along with an old ruler graduated in hundredths of an inch, but the ruler has been used so much that the ends have been worn very considerably. He measures the 1000 beams and finds a range of depths between 17.76 and 18.01 inches. It is seen that, in addition to the inherent random error, an error that always has the same algebraic sign (in this case about 0.11 inch) has been inserted. Such an error is called a systematic error.

If the systematic error is always of constant magnitude, it merely shifts the entire range of values either up or down the scale. If it changes in magnitude during the course of the experiment, the relation of the measurements, one to another, is altered and little can be said. In the limit, as the changes become more and more chaotic, systematic error may be considered random.

Other examples of systematic error would be:

1. Improper bonding of electric resistance strain gage.
2. Support which offers moment restraint when a hinge is desired.
3. Incorrect calibration of a measuring instrument.
4. Use of radial pressure in place of vertical pressure.
5. Effect of unknown residual stresses on the buckling of a compression element.

3.3 Statistics of Measurements

A rather extensive mathematical theory has been formulated which enables the engineer to make logical quantitative statements concerning the behavior of a structural system which is influenced by random fluctuations. It has already been stated that many of the errors involved in an experimental small scale model study may be of a systematic nature and hence the model results may not be amenable to statistical argument.

Nevertheless, there are many experimental phenomena which are random, and the model engineer should certainly be aware of the basic techniques for the statistical treatment of random phenomenon. More complete treatments of this subject will be found in numerous books and papers.[4,36,37,46]

3.3.1 Measurements in Model Testing. Almost all of the experimental data that are collected in a structural models laboratory are described by a numerical magnitude and as such are discrete. On the other hand, structural engineering problems seldom encompass situations wherein the possible outcomes are discretely distributed. For example, if one were making a measurement of buckling pressure on some particular thin-shell model, that pressure is not by nature restricted to have a magnitude equal to some one of a number of discrete values. It may be that the measuring system is capable only of determining from among the discrete values such as 0.254, 0.255, 0.256, 0.257 psi, for example, but in fact the actual magnitude may have been closer to 0.2554327 or 0.255432756789 or even 0.2554327568927427842784, etc. Such problems lie in the domain of a continuously distributed variable.

Table 3.2. Measurements of Buckling Pressure on Twenty Thin-Shell Models

Trial	Measured Pressure	Trial	Measured Pressure
1	0.255	11	0.250
2	0.250	12	0.245
3	0.240	13	0.255
4	0.260	14	0.235
5	0.255	15	0.250
6	0.250	16	0.250
7	0.240	17	0.255
8	0.250	18	0.240
9	0.245	19	0.260
10	0.270	20	0.245

If one were to manufacture twenty seemingly identical thin-shell plastic models and then measure a buckling pressure on each model, one would not always obtain the same buckling pressure. A typical set of measurements might be as indicated in Table 3.2. This table can easily be transformed into a histogram as shown in Figure 3.1 where the block type of diagram is used to indicate that the measurements came from a continuous system. It is of considerable use to be able to quantitatively describe such a series of measurements with as few terms as possible. Various possibilities exist, but the most common and

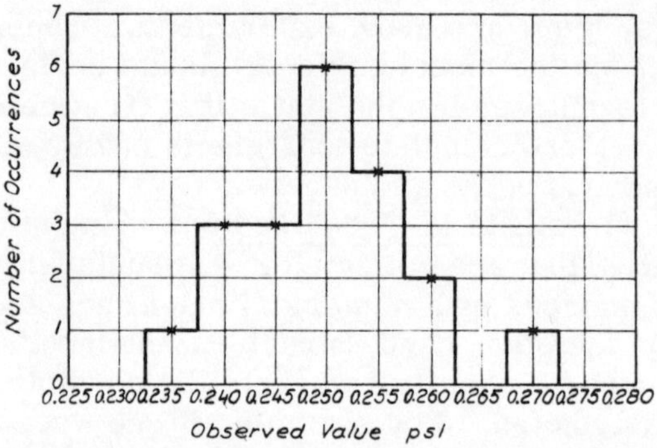

Figure 3.1. Histogram: observations of buckling pressures.

most useful consists of computing the mean value and the variance. The mean value is determined as

$$\bar{x} = \frac{\sum_{i=1}^{n} x_i}{n} \tag{3.1}$$

and provides a measure of central tendency. The variance is defined to be the second moment about the mean. Thus the variance

$$s^2 = \frac{\sum_{i=1}^{n} (x_i - \bar{x})^2}{n} \tag{3.2}$$

provides a quantitative measure of the dispersion of the measurements about the mean value. The standard deviation is merely the square root of the variance.

$$s = \sqrt{\frac{\sum_{i=1}^{n} (x_i - \bar{x})^2}{n}} \tag{3.3}$$

From these definitions one can compute for the measurements of Table 3.2:

$$\bar{x} = 0.250 \text{ psi}$$

$$s = 0.00806 \text{ psi}$$

If one were to make more and more models and at the same time were able to continuously reduce the magnitude of the smallest interval (here 0.005 psi), the histogram of Figure 3.1 would approach some kind of a smooth curve. The exact nature of this continuous curve depends upon the process which is generating the measurements. It is at this point that the experimental model engineer encounters a real difficulty. Seldom will he want to make a large number of measurements or tests; however, this very lack of a large number of tests makes it difficult to predict the true probability density function from which the measurements are being drawn. It has been found that many of the experimental measurements in science and engineering seem to approximate rather well a law known as the normal probability density function. If it can be reasonably assumed that a set of measurements has come from a population governed by the normal probability density function, then many statistical inferences can be drawn with regard to the phenomenon. On the other hand, it may be very difficult to establish the likelihood of similarity between an observed set of measurements and a normal probability density function. In these cases, it may be useful to employ certain known facts which are valid regardless of the statistical distribution.

It is hoped that the following two questions can be answered: (1) Of course the mean value $\overline{x} = 0.250$ psi of the twenty measurements is merely the mean value of the small sample of measurements that were drawn from all possible measurements. What can be said about the relation of the mean of all possible measurements (population mean) to $\overline{x}$, and similarly, what about the relation of the population standard deviation to s? (2) If a 21st measurement were to be taken, could its magnitude be predicted? To answer these questions one can proceed from two points of view.

First, it may be desirable to be as specific as possible when extrapolating the finite set of sample measurements to predict the nature of the universe. In this case it is desired to select specific values of population mean and standard deviation which are "best." No truly satisfactory procedure has been devised for accomplishing this purpose, since at some stage it naturally becomes necessary to define what "best" implies. Ordinarily any procedure for determing what is "best" is required to yield the true value when the sample size n increases without limit and to yield the true value on the average when the number of samples of size n is made large. The result most generally agreed upon states that the "best" values of the mean and standard deviation of the entire population or universe of measurements are given by

$$\mu \approx \overline{x} \tag{3.4}$$

$$\sigma \approx s' = \sqrt{\frac{n}{n-1}}\, s \tag{3.5}$$

With these values of μ and σ, it is possible to determine, for any desired probability level, the range within which the 21st measurement would lie. First, of course, one must make some assumption regarding the nature of the probability density function which is generating the measurements. If the number of measurements is large, certain useful tests exist for determining whether the measurements are likely to have come from a normal density function; however, such is rarely the case in experimental design programs. Ordinarily one must simply assume that the density function is normal or is not normal depending perhaps on the shape of the histogram. If normality is taken, Table 3.3 permits the determination, for any desired probability level, of the range within which the 21st measurement would lie. With 95% probability

$$x_{21} = \overline{x} \pm 1.96\sqrt{\frac{\Sigma(x_i - \overline{x})^2}{n - 1}} = 0.250 \pm 0.016 \text{ psi}$$

If one is not willing to make the normality assumption, then it is always possible to use Chebyshev's inequality. It is valid regardless of the governing probability density function and states that

$$\text{Probability}\,[\,|x - \mu| \leq h\sigma\,] \geq 1 - \frac{1}{h^2} \tag{3.6}$$

where h is any constant greater than 1. For 95% probability

$$\text{Probability}\,[\,|x_{21} - \mu| \leq h\,(0.00827)] \geq 1 - \frac{1}{h^2} = 0.95$$

Thus $h = 4.47$ and

$$x_{21} = 0.250 \pm 4.47\,(0.00827) = 0.250 \pm 0.037 \text{ psi}$$

It is seen that the bounds given by the Chebyshev inequality are more conservative than the bounds which were obtained in accordance with the assumption that the buckling pressures were governed by the normal density function. Figure 3.2 indicates the over-all nature of the conservatism afforded by the use of the Chebyshev inequality and in particular shows the situation just considered.

When the sample size is small, one can intuitively imagine that an extra measurement could considerably alter the "best" estimates of μ and σ, since $\overline{x}$ and s' can vary rather significantly for small n. Since the model engineer will perhaps most often be concerned with small samples, the use of "best" estimates should be regarded with skepticism.

A second alternative which can be pursued in attempting to answer the two questions employs the concept of confidence limits (or confidence intervals). As an introduction, suppose that the

Table 3.3. Area Under Normal Density Function Curve

$$\Psi(y_1) = \frac{1}{\sqrt{2\pi}} \int_{-y_1}^{y_1} e^{-\frac{y^2}{2}} \, dy$$

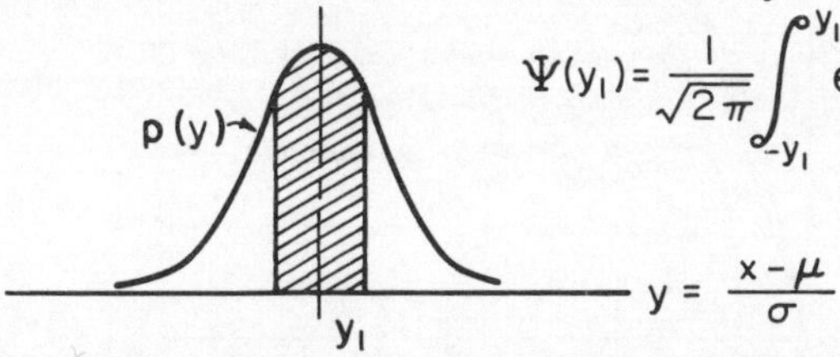

y_1	0.00	0.01	0.02	0.03	0.04	0.05	0.06	0.07	0.08	0.09
0.0	.000	.008	.016	.024	.032	.040	.048	.056	.064	.072
0.1	.080	.088	.096	.103	.111	.119	.127	.135	.143	.151
0.2	.159	.166	.174	.182	.190	.197	.205	.213	.221	.228
0.3	.236	.243	.251	.259	.266	.274	.281	.289	.296	.303
0.4	.311	.318	.326	.333	.340	.347	.354	.362	.369	.376
0.5	.383	.390	.397	.404	.411	.418	.425	.431	.438	.445
0.6	.451	.458	.465	.471	.478	.484	.491	.497	.503	.510
0.7	.516	.522	.528	.535	.541	.547	.553	.559	.565	.570
0.8	.576	.582	.588	.593	.599	.605	.610	.616	.621	.627
0.9	.632	.637	.642	.648	.653	.658	.663	.668	.673	.678
1.0	.683	.688	.692	.697	.702	.706	.711	.715	.720	.724
1.1	.729	.733	.737	.742	.746	.750	.754	.758	.762	.766
1.2	.770	.774	.778	.781	.785	.789	.792	.796	.799	.803
1.3	.806	.810	.813	.816	.820	.823	.826	.829	.832	.835
1.4	.838	.841	.844	.847	.850	.853	.856	.858	.861	.864
1.5	.866	.869	.871	.874	.876	.879	.881	.884	.886	.888
1.6	.890	.893	.895	.897	.899	.901	.903	.905	.907	.909
1.7	.911	.913	.915	.916	.918	.920	.922	.923	.925	.927
1.8	.928	.930	.931	.933	.934	.936	.937	.939	.940	.941
1.9	.943	.944	.945	.946	.948	.949	.950	.951	.952	.953
2.0	.954	.956	.957	.958	.959	.960	.961	.962	.962	.963
2.1	.964	.965	.966	.967	.968	.968	.969	.970	.971	.971
2.2	.972	.973	.974	.974	.975	.976	.976	.977	.977	.978
2.3	.979	.979	.980	.980	.981	.981	.982	.982	.983	.983
2.4	.984	.984	.984	.985	.985	.986	.986	.986	.987	.987
2.5	.988	.988	.988	.989	.989	.989	.990	.990	.990	.990
2.6	.991	.991	.991	.991	.992	.992	.992	.992	.993	.993
2.7	.993	.993	.993	.994	.994	.994	.994	.994	.995	.995
2.8	.995	.995	.995	.995	.995	.996	.996	.996	.996	.996
2.9	.996	.996	.996	.997	.997	.997	.997	.997	.997	.997
3.0	.997	.997	.997	.998	.998	.998	.998	.998	.998	.998

assumption of normality is taken and that one had prior knowledge of σ. Then using Table 3.3 it would be possible to determine limits within which a single measurement would fall as a

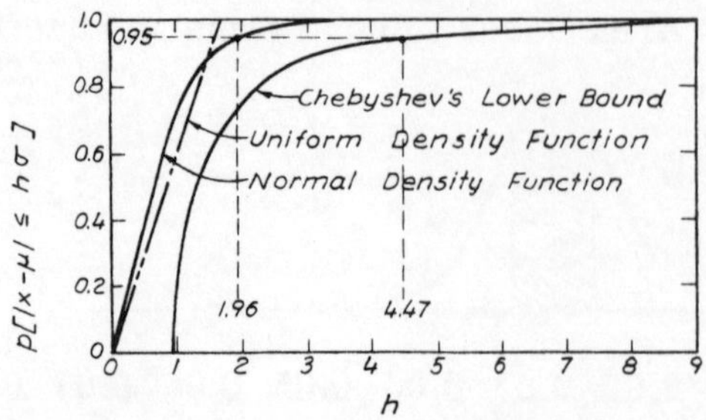

Figure 3.2. Chebyshev's inequality compared to other density functions.[37]

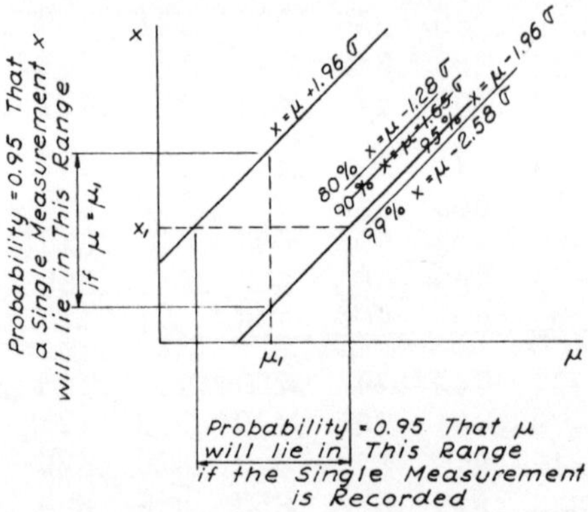

Figure 3.3. Confidence limits on μ from a single measurement x when σ is known in advance.

function of the mean value μ. For example, with a probability of 0.95, a single measurement will fall within $\mu \pm 1.96\ \sigma$, regardless of the value of μ. In view of the linearity of this relationship it must also then be true that the true mean must lie within $x \pm 1.96\ \sigma$ where x is the value of a single measurement. Figure 3.3 describes the preceding phenomenon graphically. This result can be extended to the case when the sample consists not of a single measurement but rather of n measurements.

$$\text{single measurement} \quad x - k_1\sigma < \mu < x + k_1\sigma \qquad (3.7)$$

$$\text{n measurements} \quad \bar{x} - \frac{k_1\sigma}{\sqrt{n}} < \mu < \bar{x} + \frac{k_1\sigma}{\sqrt{n}} \qquad (3.8)$$

One seldom knows the standard deviation σ, however, and consequently Equations 3.7 and 3.8 may be of little use. In this situation a single measurement cannot be interpreted; however, two or more measurements will show some spread about a sample mean. This spread can be interpreted in terms of a standard deviation s that must bear some relationship to the unknown σ. It has been shown in various statistical books that confidence limits in the aforesaid case can be given by

$$\text{n measurements} \qquad \bar{x} - \frac{ks'}{\sqrt{n}} < \mu < \bar{x} + \frac{ks'}{\sqrt{n}} \tag{3.9}$$

where values for k are given in Table 3.4 and s′ is determined according to

$$s' = \sqrt{\frac{\sum_{i=1}^{n} (x_i - \bar{x})^2}{n - 1}} \tag{3.10}$$

Table 3.4. Coefficients k for Use in Calculating Confidence Limits[46]

Number of Observations	Confidence Level 0.50	0.80	0.90	0.95	0.99
2	1.00	3.08	6.31	12.71	63.66
3	0.82	1.89	2.92	4.30	9.93
4	0.77	1.64	2.35	3.18	5.84
5	0.74	1.53	2.13	2.78	4.60
6	0.73	1.48	2.02	2.57	4.03
7	0.72	1.44	1.94	2.45	3.71
8	0.71	1.42	1.90	2.37	3.50
9	0.71	1.40	1.86	2.31	3.36
10	0.70	1.38	1.83	2.26	3.25
11	0.70	1.37	1.81	2.23	3.17
12	0.70	1.36	1.80	2.20	3.11
13	0.70	1.36	1.78	2.18	3.06
14	0.69	1.35	1.77	2.16	3.01
15	0.69	1.35	1.76	2.15	2.98
16	0.69	1.34	1.75	2.13	2.95
-	-	-	-	-	-
∞	0.67	1.28	1.65	1.96	2.58

It should be noted in Table 3.4 that the values of k listed for $n = \infty$ are in fact the k_1 values which would appear in Equations 3.7 and 3.8.

Equation 3.9 permits one to specify a confidence limit for the universe mean as opposed to the "best" estimate procedure which yielded only a single value. Confidence limits for the 21st measurement may be obtained by adding the limits from the "best" estimate procedure to the upper and lower ends of the limiting range just determined for the mean value. As applied to the 21st measurement, one obtains for 95% confidence

$$x_{21} = \left[\bar{x} \pm \frac{2.07(0.00827)}{\sqrt{20}}\right] \pm 0.016 = 0.250 \pm 0.020 \text{ psi}$$

One could go even further and determine confidence limits for the standard deviation also. If this were done and if the largest value were used in place of the "best" estimate, the limits for the 21st measurement would be widened.

It is also possible to take into account, through the Chebyshev inequality, the fact that the universe mean may not be the same as the sample mean. This is done by considering the sample mean $\bar{x}$ as a random variable and setting

$$\text{Probability } [\,|\bar{x} - \mu| \leq b] \geq 1 - \left(\frac{s_m}{b}\right)^2 \qquad (3.11)$$

where

$$s_m = \frac{s'}{\sqrt{n}} = \text{standard deviation of the mean } \bar{x}.$$

$$b = \text{some positive constant greater than } s_m.$$

The right-hand side can be expressed in terms of s', and it is then possible to determine, for any desired probability, absolute bounds on the universe mean. Of course, this additional variation spreads the final bounds in relation to those obtained by the "best estimate" approach. Thus

$$1 - \frac{1}{n}\left(\frac{s'}{b}\right)^2 = 1 - \frac{1}{20}\left(\frac{0.00827}{b}\right)^2 = 0.95$$

$$b = 0.00827$$

$$x_{21} = [\bar{x} \pm b] \pm 0.037 = 0.250 \pm 0.045$$

The Chebyshev inequality provides a conservative bound. The magnitude of this conservatism increases sharply for high confidence levels.

In summary

<u>"Best estimate" approach</u> at 95% probability

x_{21} (normal) $= 0.250 \pm 0.016$ psi

x_{21} (Chebyshev) $= 0.250 \pm 0.037$

<u>95% confidence level approach</u>

x_{21} (normal) $= 0.250 \pm 0.020$

x_{21} (Chebyshev) $= 0.250 \pm 0.045$

3.4 Propagation of Random Errors

In Section 3.3 the concern was with the statistics of measurements, that is, with determing the nature of the probability density function from which the sample of measurements was drawn and further with determining in a probabilistic way the likely outcome of an additional, as yet unmeasured, event. There are many cases, however, in which it is not enough merely to know something about the measured phenomenon. For example, if one wanted to measure the modulus of elasticity of a certain material, a customary procedure would be to measure the load and corresponding strain on a tensile specimen. From the load, the strain, and the cross-sectional area,

$$E = \frac{P}{\epsilon bt} \tag{3.12}$$

The question now arises as to whether it is possible to determine statistical relationships regarding E, where in fact there are no measurements of the quantity itself. Of course, the intent is to answer this question for a much more general class of situations than the simple product relationship in Equation 3.12.

Suppose that one has a derived quantity that is related to the directly measured values of several random variables. The functional relationship might have the general form

$$V = f(X_1, X_2, \ldots, X_n) \tag{3.13}$$

It is shown in textbooks that, if $X_1, X_2, \ldots, X_n$ are independent random variables which are closely distributed about their mean values, then

$$\overline{V} \approx f(\overline{X}_1, \overline{X}_2, \ldots, \overline{X}_n) \tag{3.14a}$$

$$s_v = \sqrt{\sum_{i=1}^{n} \left(\frac{\partial v}{\partial X_i}\right)^2 s_{X_i}^{\,2}} \tag{3.14b}$$

where the partial derivatives in Equation 3.14b are to be evaluated at the mean values $\overline{X}_1, \overline{X}_2, \ldots, \overline{X}_n$ and consequently are constants. It should be noted that the derivation which led to Equations 3.14 does not require a specification of the probability density functions of the independent random variables. However, having the knowledge of the mean and standard derivation of the derived variable does not imply knowledge of the probability density function of the derived variable even when the density functions of $X_1, X_2, \ldots, X_n$ are known. If one wants to know this additional information, then one must resort to the use of convolutions, or generating functions, or other less elementary techniques of the theory of probability. It may be stated here that, if each of the independent random variables X_1, X_2, . . . , X_n are normally distributed, then it is true that the derived random variable of a sum or difference of X_1, X_2, . . . , X_n is also normally distributed. A similar statement cannot be made when the derived variable is a product, logarithm, square root, etc., of the variables X_1, X_2, . . . , X_n. On the other hand, it is always possible to fall back on Chebyshev's inequality when one cannot easily determine the exact nature of the probability density function of the derived variable.

3.5 Example

Suppose that it is desired to check the design of a slender prismatic column. The column, which is to be of a linearly elastic material, is subjected to a compressive force P and it is of interest to determine the buckling load.

3.5.1 Analytical Solution. The mathematical formulation of this problem is well known to be expressible as

$$P_{cr} = \frac{kEI}{\ell^2} \tag{3.15}$$

It is seen that the buckling load is a function of three explicit variables and the constant k depends upon the end restraint. If these variables are considered to be random variables with known means and known standard deviations, it would be possible to determine the mean and standard deviation of P_{cr} by utilizing Equations 3.14. Of course, it would be highly desirable to be able to readily obtain the probability density function of P_{cr}, but such information cannot be obtained from the theory of error propagation embodied in Equations 3.14. Thus Equations 3.14 can be used in this example problem even if the random variables E, I, and ℓ each have different types of probability density functions, but the equations yield no information regarding the specific probability density function of P_{cr} even in the case where E, I, and ℓ all are normally distributed. With these limitations in mind one can write

$$\overline{P}_{cr} \approx \frac{k\overline{E}\overline{I}}{\overline{\ell}^2} \tag{3.16a}$$

$$\sigma_{(P_{cr})} = \sqrt{\left(\frac{k\overline{I}}{\overline{\ell}^2}\right)^2 \sigma_E^2 + \left(\frac{k\overline{E}}{\overline{\ell}^2}\right)^2 \sigma_I^2 + \left(\frac{2k\overline{E}\overline{I}}{\overline{\ell}^3}\right)^2 \sigma_\ell^2} \tag{3.16b}$$

A quantitative probabilistic prediction regarding the critical load could now be made through the use of the Chebyshev inequality.

3.5.2 Experimental Solution. If one were to undertake an experimental investigation on a small-scale structural model column as a means of determing the critical load, a dimensional analysis of the problem should be performed first. It is known that P_{cr}, E, I, and ℓ are pertinent physical variables, whereas the value of the constant k depends upon the rotational and translational end restraint (T and F having dimensions of force times length/radian and force/length, respectively). Consequently, the problem solution must be in the form

$$F(P_{cr}, E, I, \ell, T, F) = 0 \tag{3.17}$$

By selecting E and ℓ as the dimensionally independent quantities, Equation 3.17 can, according to Buckingham's theorem, be reduced to

$$\Phi\left(\frac{P_{cr}}{E\ell^2}, \frac{I}{\ell^4}, \frac{T}{E\ell^3}, \frac{F}{E\ell}\right) = 0$$

or in the solved form to

$$P_{cr} = E\ell^2\phi\left(\frac{I}{\ell^4}, \frac{T}{E\ell^3}, \frac{F}{E\ell}\right)$$

Of course, in this simple problem it is known that

$$\phi\left(\frac{I}{\ell^4}, \frac{T}{E\ell^3}, \frac{F}{E\ell}\right) = k\frac{I}{\ell^4}$$

However, such information cannot be obtained from the dimensional analysis alone. Thus the model engineer may freely select the model column length and the modulus of elasticity of the column material. Then the model restrictions and model to prototype extrapolation are given by

$$\begin{aligned}\left(\frac{I}{\ell^4}\right)_{model} &= \left(\frac{I}{\ell^4}\right)_{prototype}\\ \left(\frac{T}{E\ell^3}\right)_{model} &= \left(\frac{T}{E\ell^3}\right)_{prototype}\\ \left(\frac{F}{E\ell}\right)_{model} &= \left(\frac{F}{E\ell}\right)_{prototype}\end{aligned} \tag{3.18}$$

$$(P_{cr})_{prototype} = (P_{cr})_{model} \frac{E_{prototype}\,\ell^2_{prototype}}{E_{model}\,\ell^2_{model}} \tag{3.19}$$

Suppose that a polyvinyl chloride plastic model is carefully constructed according to Equations 3.18. If several tests (say three) are conducted, there will be some variation between the individual measurements. This variation is certainly not due to variations in I or ℓ, but could come about because of changes in E due to random temperature or relative humidity changes or because of slight changes in the end conditions, etc. If a second model were constructed, which as nearly as possible duplicated the first model, three new test results would again show some dispersion about a mean value, but in all likelihood the mean value would not be the same as the mean value obtained from the first model column. If this procedure were repeated ten times a possible set of results might be as indicated in Figure 3.4.

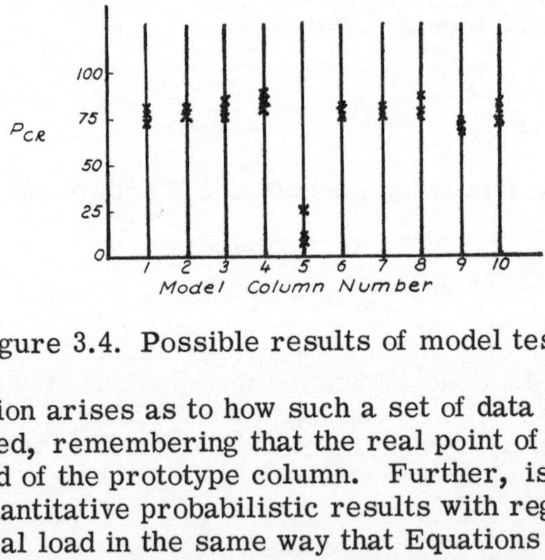

Figure 3.4. Possible results of model tests.

The question arises as to how such a set of data can and should be interpreted, remembering that the real point of interest is the buckling load of the prototype column. Further, is it possible to deduce quantitative probabilistic results with regard to the prototype critical load in the same way that Equations 3.16 and Chebyshev's inequality allowed in the analytical solution?

First, it is apparent that the results from model number five must be influenced by some blunder or systematic error. One must be very cautious about rejecting data, but it is assumed here that this model should be rejected. It is felt that some of the variation within any one of the remaining nine models may be due to errors in the measuring system (which have no counterpart in the prototype), and a more realistic determination of $(P_{cr})_{model}$ could be obtained by taking the mean and standard deviation of the individual mean values of the nine acceptable models. Of course, there may be a systematic error present in all models which causes an equal error throughout. Such an error cannot be suspected merely by inspection of the experimental data.

Now it should be noted that Equation 3.19 was deduced by setting

$$\left(\frac{P_{cr}}{E\ell^2}\right)_{prototype} = \left(\frac{P_{cr}}{E\ell^2}\right)_{model}$$

But, in light of the equalities of Equations 3.18, Equation 3.19 is certainly not a unique extrapolation equation. It could be written in several ways. For example,

$$(P_{cr})_p = (P_{cr})_m \frac{E_p\ell_p^2}{E_m\ell_m^2} = (P_{cr})_m \frac{E_pI_p\ell_m^2}{E_mI_m\ell_p^2} = (P_{cr})_m \frac{\ell_pF_p}{\ell_mF_m} = \text{etc.} \tag{3.20}$$

It is clear that all of the terms in the various right-hand sides of Equation 3.20 are in fact random variables and at this point only $(P_{cr})_{model}$ has been discussed. In this connection it must be assumed that the relevant information regarding the prototype variables E_p, I_p, ℓ_p, T_p, and F_p have been known since the start. With regard to the model variables, one can ask what particular value or magnitude of E_p, I_p, etc., is to be scaled? Should one attempt to scale the mean value of the prototype quantity, the most probable value, the value that is exceeded 99% of the time, etc. ?

If the scaling laws are applied to the mean values of the prototype and model quantities, then Equations 3.18 become

$$\left(\frac{\bar{I}}{\bar{\ell}^4}\right)_{model} = \left(\frac{\bar{I}}{\bar{\ell}^4}\right)_{prototype}$$

$$\left(\frac{\bar{T}}{\bar{E}\bar{\ell}^3}\right)_{model} = \left(\frac{\bar{T}}{\bar{E}\bar{\ell}^3}\right)_{prototype} \tag{3.21}$$

$$\left(\frac{\bar{F}}{\bar{E}\bar{\ell}}\right)_{model} = \left(\frac{\bar{F}}{\bar{E}\bar{\ell}}\right)_{prototype}$$

Now if a single value of the mean of all the quantities except $(P_{cr})_{prototype}$ were known, it is clear that if Equations 3.21 are satisfied then all of the possible extrapolation equations in Equation 3.20 will yield the same result for the mean value of $(P_{cr})_{prototype}$.

As far as obtaining a measure of the dispersion of the critical prototype load about its mean value, there is no uncertainty. In fact, it can be seen that even if exact values of the mean and standard deviation of $(P_{cr})_m$, E_m, I_m, ℓ_m, T_m, F_m, . . . , T_p, and F_p were all known, the use of Equation 3.14b with the various forms of Equation 3.20 would lead to differing values for the standard deviation of $(P_{cr})_p$. Further, none of these values would agree with Equation 3.16b, which is known to be correct. Even the second right-hand side of Equation 3.20, which would seem to be of the correct form, would not lead to a proper measure of the dispersion of the prototype critical load. This fact can be seen by considering the following hypothetical example:

<u>Prototype</u> pin-ended column

$\overline{E}$ = 29,600,000 psi $\qquad \sigma_E = 0.02\,\overline{E}$

$\bar{I}$ = 3.4 in.^{4} $\qquad \sigma_I = 0.04\,\bar{I}$

$\bar{\ell}$ = 150 in. $\qquad \sigma_\ell = 0.01\,\bar{\ell}$

<u>Model</u>

$\overline{E}$ = 450,000 psi $\qquad \sigma_E = 0.07\,\overline{E}$

$\bar{I}$ = 0.00262 in.^{4} $\qquad \sigma_I = 0.05\,\bar{I}$

$\bar{\ell}$ = 25 in. $\qquad \sigma_\ell = 0.02\,\bar{\ell}$

From Equations 3.16

$(\overline{P}_{cr})_{prototype}$ = 44,100 pounds

$\sigma(P_{cr})_{prototype}$ = 2160 pounds

On the assumption that thousands of models are tested and that the experimental technique introduces no errors except the random ones concerning E, I, and ℓ, the mean value and standard deviation of $(P_{cr})_{model}$ will also be given by Equations 3.16. Thus

$(\overline{P}_{cr})_{model}$ = 18.6 pounds

$\sigma(P_{cr})_{model}$ = 1.64 pounds

Of the various possible forms of Equation 3.20, the second would seem to be the most proper. Thus

$$(P_{cr})_{prototype} = (P_{cr})_{model} \frac{E_p I_p \ell_m^2}{E_m I_m \ell_p^2}$$

Using this relationship in conjunction with Equations 3.14, and ignoring that $(P_{cr})_{model}$ is not statistically independent of E_m, I_m, and ℓ_m, one obtains

$$(\overline{P}_{cr})prototype = 44{,}100 \text{ pounds}$$

$$\sigma(P_{cr})prototype = 5000 \text{ pounds}$$

and it is seen that the standard deviation is more than twice what is known to be correct. If only a few tests had been made and they had been affected by other experimental errors, the discrepancy would have been even larger. Since the notion of a confidence limit for the mean value of any measured quantity depends upon a knowledge of the standard deviation, such confidence limits for the mean value of the prototype critical load are subject to the lack of knowledge about the standard deviation.

3.6 Conclusions Regarding the Use of Probability Theory

It has been shown that while experimental results obtained from structural models can be used to predict something about the average or mean value of a particular physical quantity in a prototype structure, it is in general not possible to determine correct results regarding the possible dispersion about this derived mean value. Thus from one's "best" estimate of the mean value of the various physical quantities, a "best" estimate of the mean value of the prototype quantity of interest can be found. Since a correct, or in most cases even an approximately correct, value for the standard deviation of the prototype quantity cannot be obtained, one cannot make reliable probabilistic predictions regarding the outcome of the one prototype structure which is going to be built.

Ideally, it would be useful to utilize both the mean values and the standard deviations of the measured quantities in the model. Using these two pieces of information, confidence limits on the mean value of the quantity could be determined for each model quantity. When the engineer satisfies himself (that is, makes sufficient tests) that these limits are all sufficiently small, then the "best estimate" mean values should be used in the extrapolation equation.

Of course, all of the discussion in Section 3.3 through 3.5 has rather assumed that at least 3 or 5 or 15 tests have been conducted. Although such a number of tests is certainly to be desired, perhaps the most serious immediate problem facing the experimental designer is that the time and cost involved in even a single model may be prohibitive. What if in the column buckling problem only one model had been tested? What if it had been column model number 5? It has been shown that the only advantage of fabricating and testing more than one model is to obtain a better approximation of the true model mean. If the mean is taken to be the result obtained from a single model, then it is particularly important that the investigator be convinced of the absence of blunders and major systematic errors in that single model study. Such systematic errors can enter into the model results through a variety of means; for example, through the physical means of providing for the boundary supports, through incomplete bonding of a strain gage, through cementing a strain gage to a plastic material that is not resistant to the solvents in the cement, through switching circuits incorporating large switching resistances, through battery decay in a recording device, through the use of radially applied loads in place of actual gravity loads, etc. Blunder and systematic error are sometimes indicated by trends in the data, jumps in the data, a periodicity in the data, or changes in precision of the data; however, when only one or two models are to be studied, such indicators can seldom be used. Some systematic deviations from the proper course are knowingly permitted, and occasionally the effects of these deviations can be predicted mathematically. For example, it is seen in Figure 2.15 that the assumption of linearly elastic behavior for the instability of a concrete thin-shell roof may not be completely reasonable. Pahl[35] has developed a simple mathematical technique with which one can obtain a bound on the error that could be introduced by the assumption of linearity. On the other hand, many systematic error sources may be difficult to find. All possible partial checks should be taken to uncover systematic errors. For example, it may be possible to provide for several static checks within a model. Also, points of known symmetry and antisymmetry should be checked. Finally, it would be extremely useful to know that an extensive set of tests had been successfully carried out on a similar problem using techniques of fabrication, loading, and instrumentation similar to those proposed. Thus, the great background of experience underlying the use of electrical resistance strain gages on metallic materials leads one to have confidence in such results, whereas results obtained on foamed plastic materials might be suspect.

Although the example problem of the previous section was concerned with column buckling, the conclusions regarding model to prototype extrapolation are equally valid for the thin-shell buckling problem. The objective of the experimental program of this monograph is directed to the question of whether or not reliable model results can be obtained in thin-shell buckling problems. Sources of systematic error will receive the most attention, for all the statistical and probabilistic operations that have been discussed in this chapter are meaningless if the model result is in error by a factor of two or three due to some gross error.

Chapter 4

EXPERIMENTAL INVESTIGATION

The experimental program described in this chapter consisted of:

1. Fabricating twenty models of an 18-inch radius spherical cap from six different molds, and buckling these models using air pressure loading. Two types of edge restraint were employed and compared.
2. Buckling four of the 18-inch models with a vertical discrete weight loading. The load grid spacing and edge restraint conditions were varied.
3. Fabricating four models of a 36-inch radius spherical cap from two different molds, and buckling these models using air pressure loading.
4. Making various tests to establish the modulus of elasticity and Poisson's ratio of Boltaron 6200 PVC under a variety of pertinent conditions.

The results of these experiments do provide a basis for answering the questions posed in Section 2.1.

4.1 Properties of Boltaron 6200 PVC

The uniaxial stress-strain curve for Boltaron 6200 PVC, as indicated by the manufacturer, is shown in Figure 2.11. As can be seen in Figure 2.11, the tension and compression behavior of PVC are not alike even to the extent of having different tensile and compressive initial moduli of elasticity. From such considerations it can be imagined that the modulus in bending will be still different, and, in fact, that is the case. Further, the precise value of the modulus corresponding to each of the three stress conditions depends upon the stress level, temperature level, relative humidity, previous strain history, previous temperature history, rate of loading, age, etc. In view of the very large number of parameters involved, it was not clear at the start of the experimental investigation whether it would be possible to state with assurance that the buckling behavior of the model depended upon a value of modulus of elasticity reasonably close to that which could be predicted. Plus or minus 5% is meant to be reasonable. Testing samples under a variety of conditions was required in order to be certain that variations of modulus in the models were small.

4.1.1 Testing Procedure for Determining Modulus of Elasticity. Since the modulus does vary depending upon the state of stress, one must first establish what state of stress should be considered in determining a numerical value. The bending modulus is the controlling one in a buckling study. For purposes of comparison, however, preliminary tension tests were conducted, although all control specimens from the model shells were tested in bending.

It has been explained previously that the maximum stress levels in the actual tests were expected to be about 500 psi, so one is justified in seeking an initial modulus of elasticity. Three different testing techniques were employed in the tension modulus tests, namely: (1) measuring longitudinal strain with SR-4 foil strain gages, (2) measuring longitudinal strain with a special extensometer, and (3) using a constant rate-of-cross head movement tensile tester, and then computing an average longitudinal strain over the grip length. These three techniques gave values for modulus ranging from 280,000 to 666,000 psi.

The tests utilizing electric resistance strain gages were very simple. Two 1/4-inch gages were cemented to opposite sides of 8 × 0.66 × 0.032-inch and 8 × 0.69 × 0.065-inch specimens, and the tests were run as shown in Figure 4.1. Assuming that the strain in the gages is the same as that in the plastic, one might think that the modulus of elasticity could be obtained directly from the stress-strain curves which are shown on Plates A-1 and A-2 in Appendix I. Accordingly, values of E = 666,000 psi and 596,000 psi are obtained for the 0.032 and 0.065-inch specimens, respectively. Now it happens that these two specimens were especially chosen so that their bending moduli, as determined by a cantilever beam test to be discussed later, were the same (475,000 psi). From the resulting discrepancy, from the fact that the manufacturer claims a modulus of 400 to 500,000 psi, and from the previous experience of other investigators, it is clear that the resistance strain gages significantly affect the plastic. It might be assumed that the SR-4 instrumented system behaves according to

$$P = \epsilon[A_{\text{plastic}}\, E_{\text{plastic}} + (AE)_{\text{gages}}] \tag{4.1}$$

By testing two samples with different plastic areas, two independent Equations 4.1 can be obtained. Assuming E_{plastic} to be the same for the two samples, the resulting equations can be solved for E_{plastic}, since E_{plastic} and $(AE)_{\text{gages}}$ are the only two unknowns. Such a procedure yields E = 536,000 psi for these two specimens. It is felt that this value is still too high, so that things beside simple mechanical stiffening are involved. Gage heating must be one of these things.

Other tests, such as the one reported on Plate A-9, confirm the severe influence of the SR-4 gages. Some previous investigators have attempted to evaluate the stiffening effect,[10,20]

Figure 4.1. Experimental setup for SR-4 gage tensile tests.

but their efforts have been narrow in scope. Although strains are not to be measured in the buckling model tests of this monograph, it is clear that stress distribution studies of thin plastic shells by means of SR-4 measurements may be considerably in error due just to gage stiffening and heating.

The second technique employed an electromechanical extensometer. At first, an extensometer having resistance strain gages was used; but later, the inductance-type extensometer shown in Figure 4.2 was used. Stress-strain curves in tension were obtained on six specimens, three oriented parallel to the direction of calendering during manufacture and three perpendicular. These curves are shown on Plate A-3 from which it is seen that the initial modulus is approximately 450,000 psi.

The final tensile testing technique employed the Instron Tensile Tester shown in Figure 4.3 without any gage. A constant rate-of-crosshead movement can be converted to an average strain over the grip length. A typical raw data sheet from such a test is shown in Plate A-4. It is apparent that the grip length is critical, since apparent moduli vary from about 280,000 psi for a grip length of 2 inches up to about 450,000 psi for a grip length of 20 inches. A summary of the results from six test samples is shown in Plate A-5. Assuming that the local end effect would be insignificant if a grip length of infinity were used, one could extrapolate the given results back to the infinite sample length. A modulus between 400,000 and 500,000 psi results from such an extrapolation.

As a means for obtaining the bending modulus, the deflection of a cantilever beam was measured. Figure 4.4 shows the

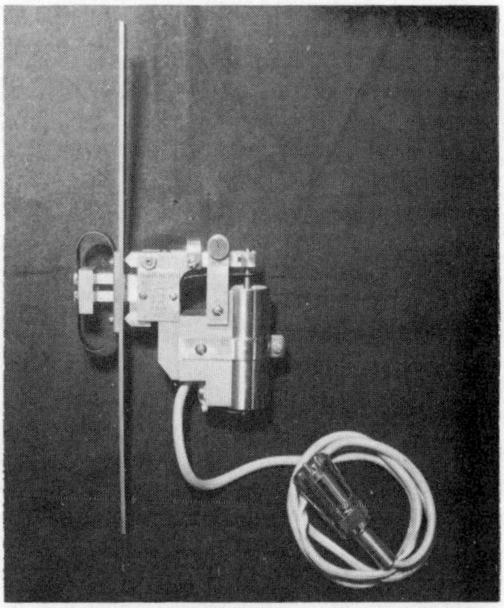

Figure 4.2. Electromechanical extensometer.

Figure 4.3. Instron Tensile Tester.

clamping of the specimens. The procedure for testing was established after considering the effects of

1. Poisson's ratio (since beams 0.03 inch thick were used,

Figure 4.4. Cantilever beam test for bending modulus.

plate action must be present).

2. Nonlinearity (when deflections exceed the thickness nonlinearity may be important).
3. End fixity.

Plate A-6 presents results of cantilever beam tests on 0.698 × 0.0304 inch, 0.372 × 0.0306 inch, and 0.296 × 0.0302 inch specimens. The smaller specimens were obtained by splitting the 0.698-inch specimen. A load was applied at 3 inches and the deflection measured at 2 inches with a 0.001-inch micrometer. A Poisson's ratio effect was observed. As a result of this test, and other similar tests, it was concluded that the modulus as determined from a specimen 0.7 ± inch wide should be multiplied by a factor of 0.96.

In order to obtain deflections that were large enough to be an order of magnitude larger than the inherent error in the micrometer measurement, the domain of large deflection is entered. Euler,[11] in his original studies of elastic curves, employed the exact curvature expression, which for the cantilever beam leads to the governing load-deflection equation

$$EI \frac{y''}{(1 + y'^2)^{3/2}} = Px \tag{4.2}$$

where

P = concentrated load

x = distance measured from concentrated load, positive in direction back toward the fixed end

y = deflection, positive downward

Δ' = end deflection

Euler integrates Equation 4.2 by series and shows that

$$6EI\Delta' = P\ell^2(2\ell - 3\Delta') \tag{4.3}$$

Of course, if the $3\Delta'$ on the right-hand side were neglected, Equation 4.3 would reduce to the well-known cantilever beam deflection formula. When deflections of 0.075 inch are induced at a distance of 2 inches, Equation 4.3 would predict that the nonlinear effect is approximately 5%, and then the computed modulus is 5% too high. Plate A-7 shows the results of tests which were conducted to investigate nonlinearity (as well as end fixity). It is seen that deflections varying from 0.20 to 0.03 inch did not appear to lead to different computed moduli. Surely Equation 4.3 must be correct, but it is important to note that in Equation 4.3 Δ' is the vertical deflection of a certain point. A horizontal movement accompanies this vertical deflection. Experimentally Δ' was not measured, but rather a deflection Δ corresponding to the vertical distance between a certain point in the undeformed beam and some other point directly below it in the deformed beam. It is necessary that Δ be larger than Δ' and, consequently, the use of Δ tends to cancel the 5% nonlinear effect predicted by Equation 4.3. For engineering purposes, it is felt that the proposed procedure is satisfactory. No compensation is made for the nonlinear effect.

Finally, it remains to consider whether absolute end fixity can be obtained. No test could be devised which would prove that full fixity was provided. Plates A-7 and A-8 and the generally consistent results obtained throughout all the control specimen testing for the model shells (see Figure 4.5) would indicate that full fixity was obtained. Henceforth, full fixity was assumed.

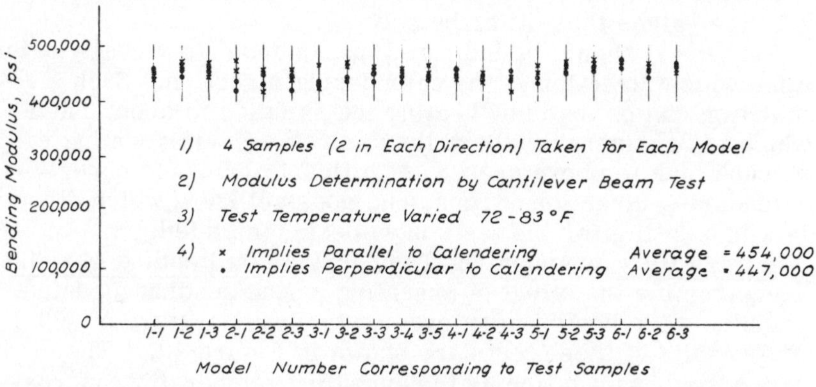

Figure 4.5. Summary of bending modulus for test samples from 18-inch radius model shells.

In summary, the bending modulus for the model control specimens was to be computed from the deflection at 2 inches on a 3-inch cantilever beam. The applied load was 5.93 and 47.1 grams for the 1/32 and 1/16-inch samples, respectively. A value of E was computed using the elementary beam theory, and then this value was multiplied by 0.96 in order to account for the Poisson's ratio effect.

It should be mentioned that the 5.93 grams, acting at 3 inches on a 0.07×0.03-inch specimen, causes a maximum stress of 445 psi, or of the same order of stress magnitude to be expected in the model just prior to buckling.

4.1.2 Summary of Modulus Test Samples Taken for Shell Models. For each shell model, four test samples were made and tested as cantilever beams. These samples were taken from the edges of the vacuum-formed sheet, two in each direction. Figure 4.5 brings all these tests on to one graph. It is apparent that the manufacturing quality control of the PVC sheets is very good.

4.1.3 Directional Properties of Manufactured Sheets. Since Boltaron 6200 PVC sheets are manufactured by a calendering process, the mechanical properties may vary with direction. The stress-strain curves obtained using the M. I. T. extensometer (Plate A-3) show a definite orientation effect, but fortunately there appears to be only about a 3% difference between the modulus of elasticity parallel to the calendering direction and that perpendicular to it. Figure 4.5 indicates the orientation effect which was noted in the cantilever beam tests. The errors in the cantilever beam tests sometimes exceed the orientation difference, so that in only 9 of the 20 model shells do both of the parallel moduli exceed the perpendicular moduli. On the other hand, the 40 parallel samples gave an average modulus of 454,000 psi compared to 447,000 psi for the 40 perpendicular samples. The average values thus differ by only 2%.

Perhaps it should be indicated that one cannot visually determine which direction is the calendering direction. Such a determination can be made by heating the samples to about 350 °F, which is considerably above the desired vacuum-forming range. At such high temperatures, a significant shrinkage occurs in the calendering direction. Thus, one does not know which direction is which until after the tests have been completed.

4.1.4 Effect of Annealing Temperature on Bending Modulus. To determine the effect of annealing on the bending modulus, samples were heated in an oven up to a temperature of 400 °F. The results of these tests are shown in Figure 4.6. They indicate a slight relaxation as the annealing temperature increases, but the change from no annealing to the extreme of 400 °F causes only a 15 to 20% decrease in the modulus. Over the vacuum-forming range, the modulus variation is only about 3%.

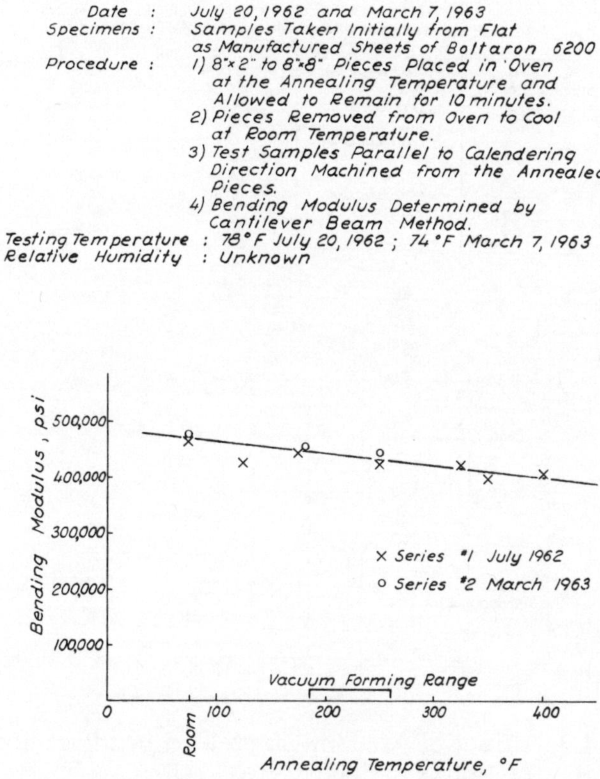

Figure 4.6. Effect of annealing temperature on bending modulus of Boltaron 6200 PVC.

4.1.5 Effect of Vacuum-Stretch on Bending Modulus. In the model domes formed over a male mold, it was observed that the plastic underwent no stretching at the top. As the plastic was pulled down over the mold, the amount of stretch increased very gradually as the distance down a meridian increased. Around the edges the stretch was nearly biaxial and approximately 12%. To investigate the possible effects of such stretching, a model was vacuum-formed over a 4 × 4 × 7-inch high block of Plexiglas. From this model, samples could be taken which had undergone various amounts and directions of stretch. The results of these tests are shown in Figure 4.7 and are self-explanatory. Applying the results to the model domes, the edges of the domes formed on a male mold and the center of those formed on the female mold may have an effective modulus of elasticity 5 to 10% lower than that existing in other parts of the shell.

The modulus test samples for the shell models were taken from the corners of the plastic sheet and therefore did not undergo any stretching.

Sample	Orientation in Preformed Sheet Re: Calendering	Linear Stretch in Longitudinal Direction	Linear Stretch in Transverse Direction
1a	Perpendicular	80 %	-15 %
1b	"	80 %	-15 %
2a	Parallel	0	0
2b	"	0	0
3a	Perpendicular	0	0
3b	"	0	0
4	Parallel	15 %	10 %
5	Perpendicular	10 %	15 %
6a	Parallel	0	80 %
6b	"	0	80 %

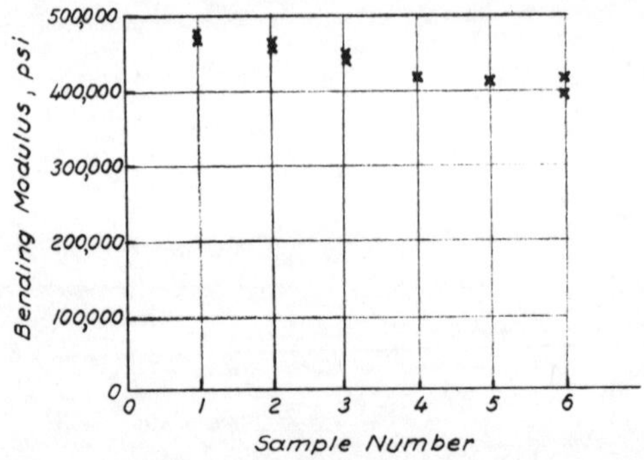

Figure 4.7. Effect of vacuum-stretch on bending modulus of Boltaron 6200 PVC.

4.1.6 Effect of Aging on Bending Modulus. To investigate whether the bending modulus might change with time, the modulus samples for model shell 4-1 were placed in the laboratory and tested from time to time over a period of 311 days. From the results in Figure 4.8, it can be concluded that for the period of time over which tests were conducted there was no significant change in the bending modulus of elasticity.

Visual observations indicate a reduction in ductility with time. However, since ductility is not important in an elastic stability study, no quantitative measurements were taken.

4.1.7 Effect of Environmental Conditions on Modulus. The dependence of modulus of elasticity on temperature which is shown in Figure 2.13 has been furnished by the manufacturer. In the 70 to 90°F temperature range the modulus of elasticity decreases at the rate of about 1000 psi/°F. The Laboratory for Structural Models in the Civil Engineering Department did not have temperature control at the time these tests were made, and so there was usually a temperature change between the time when the modulus of elasticity specimens were tested and the time when the model shells were buckled. From all temperatures recorded in the test data tables of Appendix B, it is seen that the

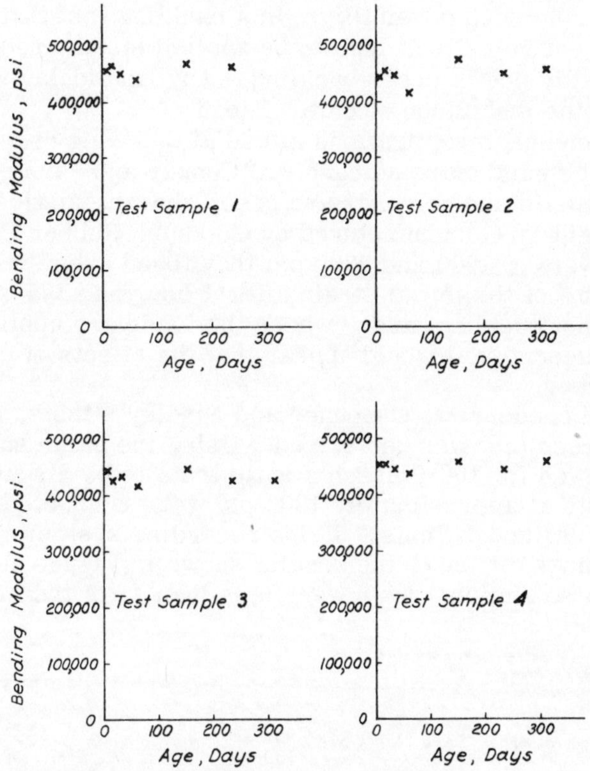

Figure 4.8. Effect of age on bending modulus of Boltaron 6200 PVC.

maximum temperature difference between modulus test and buckling test is 11°F. Even though the modification would be at most 3%, the values of modulus of elasticity obtained from the bending tests were modified by 1000 psi/°F for the difference between temperature at the time of modulus determination and that at the time of buckling test.

The effect of relative humidity on the modulus of elasticity is unknown, although it is felt to be small. There was no control of a measurement of relative humidity during the buckling tests.

The Instron Tensile Tester afforded a means of obtaining data on the effect of strain-rate upon the modulus of elasticity. No pattern of strain-rate influence upon the modulus of elasticity could be noted from all of the data leading to Plate A-5. Any effect must be small.

4.1.8 Creep Characteristics. In Section 2.1.2, some consideration has already been extended to the problem of creep strains. It is often assumed in model tests utilizing plastic materials that the plastic exhibits an effective modulus that decreases with time.[17] For tests involving strain readings, it is customary to take readings 10 or 15 minutes after load application. Buckling tests offer no such possibility. In a buckling test that does not allow the entire critical load to be applied instantaneously, the test is really one of creep buckling. Any time delay can only decrease the magnitude of critical load.

Experimental investigations aimed at deducing creep characteristics demand extreme care. O'Conner and Findley[33] have studied tensile and compressive creep characteristics of Geon 404, a rigid PVC manufactured by Goodrich Rubber Co. At stress levels of 1000 and 2000 psi they found creep strains equaling 5% of the initial strain after 1 hour and 17% after 500 hours. In a buckling study, where the loads are applied in 2 or 10 or 60 seconds, it would appear that the effects of creep are slight indeed.

To lend credence to O'Conner and Findley's tests, a crude tensile creep test was performed. Using the same specimens which served for the Poisson's ratio tests to be discussed later, creep tests at approximately 1000 psi were conducted for 200 minutes. Although "noise" in the recording system made precise readings difficult, the results shown in Figure 4.9 show that the creep strains in 1 hour were less than 4% of the initial strain.

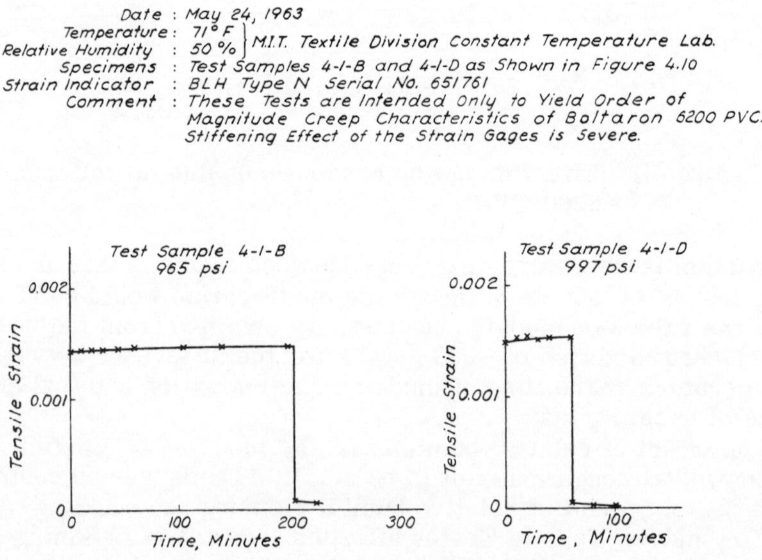

Figure 4.9. Tension creep characteristics of Boltaron 6200 PVC.

Since air pressure loads were to be applied in less than 30 seconds and weight loadings almost instantaneously, creep strains should not be important. Air pressure tests, with time to critical pressure varying from 10 to 180 seconds, showed no pattern of pressure change, so a time to buckling of 15 to 30 seconds was adopted and maintained in all air pressure tests. After the buckling pressure had been determined, it was possible to load a model to slightly less than the critical pressure and then to wait for buckling to occur. Buckling times were erratic, indicating that variables such as temperature, relative humidity, air currents, etc., would have to be very carefully controlled in order to use plastic models in studies of creep buckling.

4.1.9 Poisson's Ratio. Poisson's ratio for engineering materials varies from about 0.18 for concrete up to 0.50 for some rubber materials. Plastic materials have Poisson's ratios varying from about 0.35 to 0.50. Since Poisson's ratio is a pertinent physical quantity in an elastic buckling study, a test was conducted to obtain a reasonable value for Boltaron 6200 PVC in the 0 to 1000 psi stress range. Using specimens such as that in Figure 4.10, and a test setup such as that in Figure 4.1, the data for indicated longitudinal versus transverse strain is shown on Plate A-9. Using these results, a value of $\nu = 0.38$ was used for the interpretation of all shell model tests.

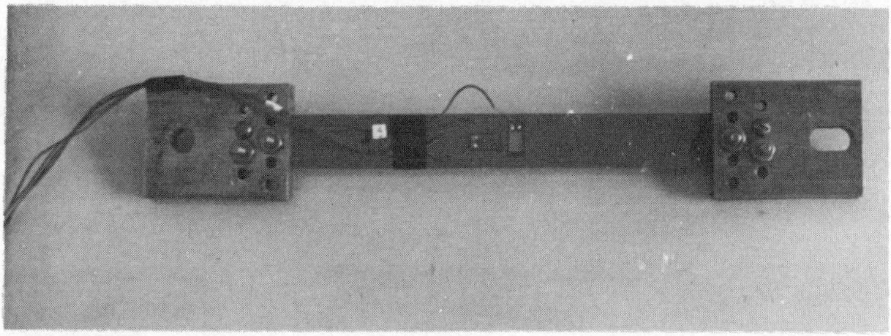

Figure 4.10. Specimens for Poisson's ratio test.

4.2 18-Inch Radius Domes Subjected to Air Pressure

Twenty shell models from six molds were fabricated and tested. The experimental procedure and results obtained are included in this section. For identification purposes, shell 3-2 indicates that the model was the second one pressed on mold No. 3.

4.2.1 Fabrication and Testing Procedure. Six molds were made for this series of tests. Mold No. 1 was of wood, with an epoxy paint coating, and was made in a pattern shop. The intention was to compare such a "professionally" finished mold with later homemade plaster molds. The plaster molds were constructed using gypsum plasters. Hydrocal A-11, Hydrocal B-11, and Ultracal 30 were used, the changes being made only because of successively better tooling characteristics. The holding base and screed are shown in Figure 4.11, and a finished mold in Figure 4.12. The finished mold is placed in a vacuum press shown in Figure 4.13. Figure 4.14 shows diagrammatically the steps involved in this vacuum-forming process. A shell model as it comes from the vacuum-forming machine is shown in Figure 4.15.

Figure 4.11. Mold base with screed for 18-inch radius mold.

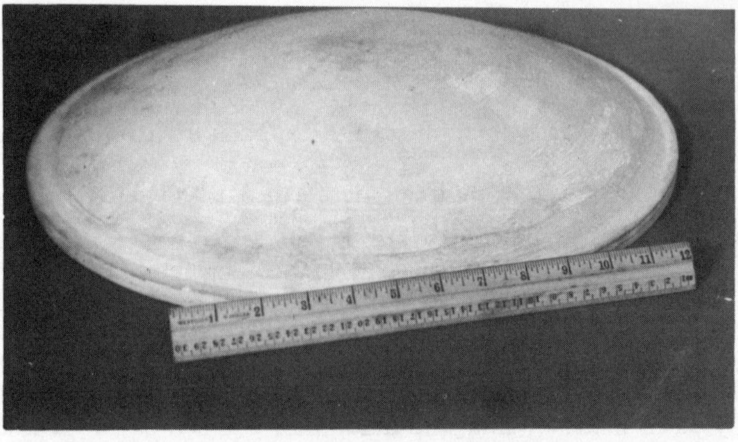

Figure 4.12. Typical finished 18-inch radius mold.

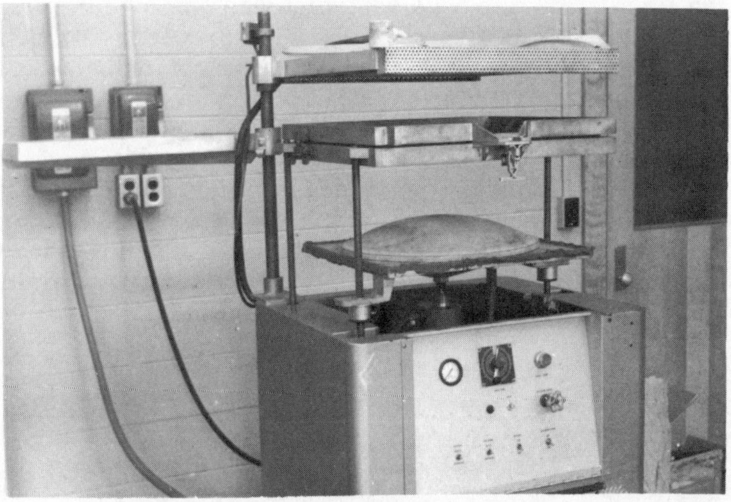

Figure 4.13. Vacuum-forming machine.

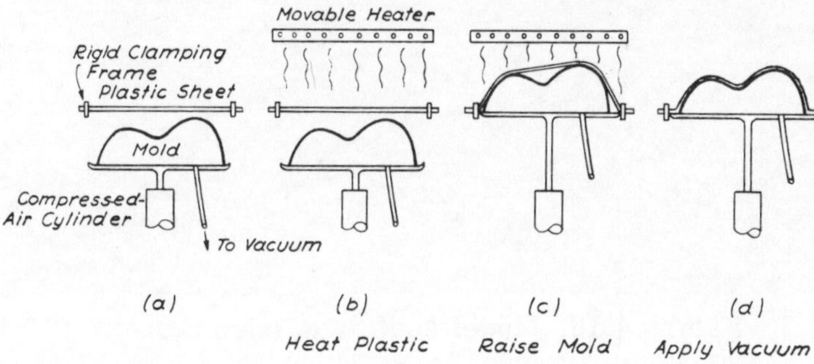

Figure 4.14. Schematic vacuum drape forming sequence.

Figure 4.15. Model shell before trimming.

In the first tests, the edge restraint was obtained by clamping the flange as shown in Figure 2.17. This procedure was later abandoned, and the edge restraint was made by embedding the shell edge in an epoxy cement (EP-F Epoxy Cement, Schwartz Chemical Co., New York). A model with the edge flange is shown in Figure 4.16, while one with the flange removed is shown in Figure 4.17. A typical cement edge after removal from the test equipment is shown in Figures 4.18 and 4.19. In order to get a quantitative measurement of the edge restraint furnished by such a cement fillet, cantilever beams were fixed in a similar manner. By comparing the deflections with those that had been

Figure 4.16. Model shell with edge flange.

Figure 4.17. Model shell without edge flange.

Figure 4.18. Epoxy cement edge condition.

Figure 4.19. Epoxy cement edge condition.

observed in the "standard" cantilever beam tests, a measure of end restraint could be determined. The samples were cemented, tested, removed, recemented, and tested again. The data are recorded on Plate A-10. It was concluded that essentially full fixity is obtained by the cementing procedure. It should be noted, however, that the epoxy was flexible, having a modulus of elasticity of perhaps 50 to 100,000 psi, as opposed to the rigid epoxies with their moduli of 500 to 700,000 psi. The one convenient fact regarding the flexible epoxy cement was that it was not a very good cement and could be removed without damaging the model shells.

Model shell thicknesses at 41 different surface points were measured with a micrometer and the Ames dial arrangement shown in Figure 2.18.

Figure 4.20. Air pressure test equipment for 18-inch radius domes.

It was noted in Figure 2.15, and will be seen again later, that the membrane compressive stresses at the point of shell buckling were of the order of 250 psi. In the theoretical Euler column buckling problem, residual stresses are of no consequence unless the sum of the residual and applied stresses exceeds the linearly elastic stress range. If such reasoning is directly applicable to the shell buckling problem, then residual stresses in the plastic model shells would have to be very large in order to have any appreciable effect. On the other hand, there is really no reason to suspect that residual stresses affect shell buckling in the same way as they affect column buckling. In order to get an impression of the possible magnitude of residual stresses in the plastic shell models, three of the tested shells were cut along a great circle, and no shape change was noticed. The half-shells fitted snugly onto the uncut shells. For a quantitative measure, SR-4 rectangular rosettes were glued to the top and bottom surfaces of shell 6-1 at the crown and at one position along the edge. The rosettes were then cut out and the strain changes were recorded. The resulting data are recorded in Plate A-11. Preliminary experimentation indicated that gage heating, cutting, and handling problems would limit the accuracy of the strain readings to ± 30 microinches/inch. Although the observed strains were of the

Figure 4.21. Model prebuckled.

Figure 4.22. Model postbuckled.

order of the inherent error, it can be said that the residual stresses are of the order of 50 psi or less. Thus, residual stresses may be of the order of 10% of the stresses induced by the buckling load.

The test equipment which was described in Figure 2.16 is shown in Figure 4.20. Just prior to testing, the geometry was checked with a template as discussed in Section 2.4. Figures 4.21 and 4.22 show a model just prior to buckling and just after buckling. No high-speed photographs of the buckling process were made—Klöppel and Jungbluth[24] and Schmidt[38] having already presented such information. Temperatures were measured at the time of each test, but no measurement was taken for relative humidity. The loading time varied between 15 and 30 seconds. For each test condition the shell was buckled, was relieved and allowed to rest for 30 minutes or longer, was buckled again, was relieved and allowed to rest for 30 minutes or longer, and was buckled again. The three buckling pressures, without exception, repeated almost identically. Then the model was removed from the testing fixture.

4.2.2 Data. A summary of factors attendant to each of the twenty model shells and the data regarding the buckling capacity of the shells is presented in Appendix B.

4.2.3 Results and Conclusions. The results from the twenty shell models are shown in Tables 4.1 to 4.3 and Figures 4.23, 4.25, 4.26, and 4.27. Table 4.1 merely summarizes some of the data. Without exception the buckle position was near the top. The location and repeatability of buckle positions are shown on the shell thickness diagrams in Appendix B.

The classical theory for the complete sphere states that p_{cr}/E is proportional to the second power of t/R. Even if one goes over to the classical theory as applied to shallow spherical caps, it is seen that for the twenty shells

$$21.6 < \lambda = \sqrt[4]{12(1-\nu^2)} \; \frac{r_0}{\sqrt{Rt}} < 24.6$$

Thus, even if one thought that such shells as are tested here could be treated as shallow, the predicted critical pressure would likely be very close to that predicted for the complete sphere.

Figure 4.23 compares the correlation of the data for the epoxy cemented edges when the shell thickness is taken to be that in the vicinity of the buckle position and to be the average thickness over the entire surface area. Theoreticians in the past[22, 44] have felt that, since the buckling of the dome occurs by a local dimpling in one isolated spot, the entire phenomenon was a local one. If this were true, the correlation of Figure 4.23 should be better for the thickness in the vicinity of the buckle. Since this correlation is poor relative to that for the average thickness, it is clear that the restraint which the remainder of the shell offers to the local buckled area is important. Of more importance to this study is the fact that Figure 4.23 sheds some light on the question of what thickness should be used in the model-prototype scaling relationships dictated by dimensional analysis. There is a danger, of course, in attempting to apply the results of this restricted study to all thin-shell buckling situations; but it does seem reasonable to conclude that, for those situations governed by so-called local buckling, the average shell thickness should be used in the scaling relationships.

Another curious fact regarding Figure 4.23 is that the data for

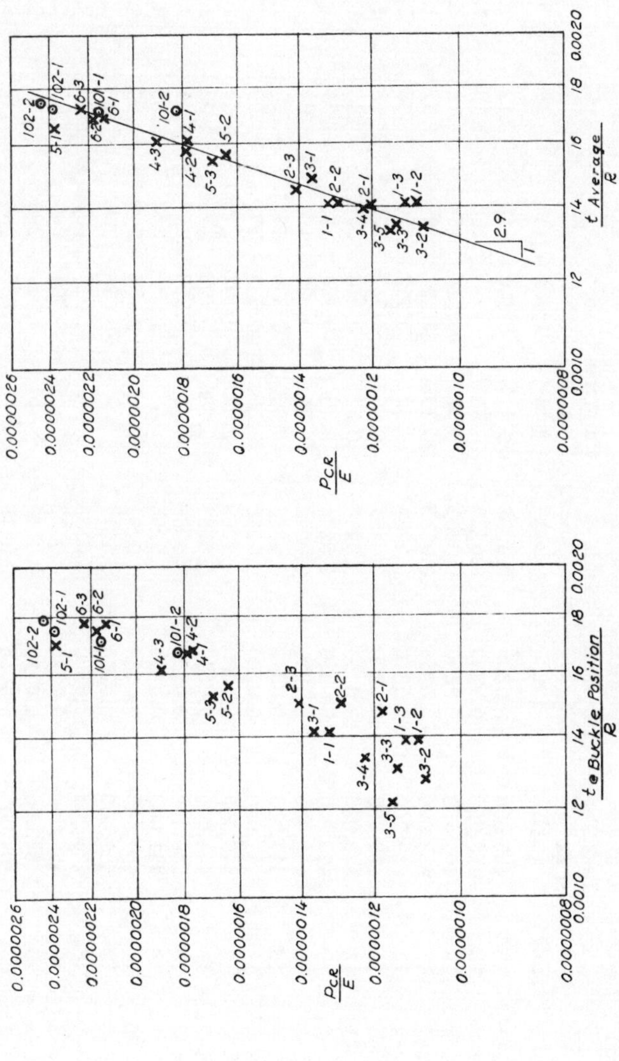

Figure 4.23. Critical pressure correlation for average and local thicknesses.

Table 4.1. Summary of Test Data for Domes Loaded by Air Pressure

Shell	Radius (in.)	Bending Modulus Corrected to Model Test Temperature (psi)	Thickness at Buckle Position for Epoxy Edge (in.)	Average Shell Thickness (in.)	p'_{cr} Clamped Flange (psi)	p_{cr} Epoxy Joint (psi)	$\frac{p_{cr} R}{2t_{ave}}$ (psi)	Classical Buckling Pressure p_{cl} (psi)	$\frac{p_{cr}}{E}$ $\times 10^{-6}$	$\frac{p_{cr}}{p_{cl}}$
1-1	18	455,000	0.0255	0.0253	0.517	0.600	213	1.12	1.32	0.54
1-2	18	478,000	0.0250	0.0253	0.521	-	186	1.18	1.09	0.44
1-3	18	461,000	0.0250	0.0252	0.523	0.514	184	1.13	1.12	0.46
2-1	18	437,000	0.0265	0.0252	0.305	0.527	188	1.07	1.20	0.49
2-2	18	435,000	0.0270	0.0254	0.327	0.563	200	1.08	1.29	0.52
2-3	18	437,000	0.0270	0.0261	0.383	0.617	213	1.15	1.41	0.54
3-1	18	449,000	0.0255	0.0267	0.510	0.610	206	1.23	1.36	0.50
3-2	18	463,000	0.0230	0.0241	0.456	0.498	186	1.04	1.08	0.48
3-3	18	458,000	0.0235	0.0242	0.345	0.522	194	1.04	1.14	0.50
3-4	18	453,000	0.0240	0.0251	0.460	0.555	199	1.10	1.22	0.50
3-5	18	461,000	0.0220	0.0239	0.393	0.533	201	1.02	1.15	0.52
4-1	18	445,000	0.0300	0.0289	0.521	0.792	247	1.43	1.78	0.55
4-2	18	444,000	0.0300	0.0284	0.486	0.793	251	1.38	1.79	0.57
4-3	18	436,000	0.0290	0.0289	0.495	0.829	258	1.41	1.90	0.59
5-1	18	451,000	0.0305	0.0298	-	1.07	323	1.55	2.37	0.69
5-2	18	460,000	0.0280	0.0281	-	0.755	242	1.40	1.64	0.54
5-3	18	469,000	0.0275	0.0277	-	0.790	257	1.39	1.68	0.57
6-1	18	475,000	0.0320	0.0304	-	1.01	299	1.70	2.13	0.60
6-2	18	459,000	0.0315	0.0303	-	0.997	296	1.63	2.17	0.61
6-3	18	457,000	0.0320	0.0310	-	1.02	296	1.69	2.23	0.60
101-1	36	407,000	0.0615	0.0616	-	0.875	256	1.49	2.15	0.59
101-2	36	415,000	0.0600	0.0617	-	0.757	220	1.52	1.83	0.50
102-1	36	411,000	0.0630	0.0621	-	0.978	283	1.53	2.38	0.64
102-2	36	417,000	0.0645	0.0629	-	1.02	292	1.59	2.44	0.64

Notes on Table 4.1

Column 3. Four test samples were taken, two in each direction, from the edge of the vacuum-formed sheet. The average value from these tests was modified by 1000 psi/°F for the difference between modulus test and model test temperatures.

Column 4. This is the thickness at the place where buckling initiated. For shell 1-2 the flange-clamped buckle position was used.

Column 5. This is an average thickness over the entire surface area, not just the average of the 41 measured points.

Column 6. This is the average of the test results for the flange-clamped edge condition. The series No. 1 results are included in Appendix B. Similar data are not given for series Nos. 2, 3, and 4, but the values given here are the average from two or more separate clampings.

Column 7. This is the average of the test results for the epoxy-glued edge condition. Shell 1-2 was not tested in this manner since it was desired to retain one model with an intact flange. For subsequent interpretation, p_{cr} was assumed equal to p'_{cr}.

Table 4.2. Repeatability for Any One Model

Shell	Date	Type of Edge Restraint	p_{cr}	Buckle Position
1-1	6/20/62	Flange-Clamped	0.49	Near Top
	6/20/62	Flange-Clamped	0.53	Near Top
	4/11/62	Flange-Clamped	0.53	Mid Height
	4/11/63	Flange-Clamped	0.48	Mid Height
	4/11/63	Flange-Clamped	0.55	Indifferent
	4/11/63	Flange-Clamped	0.52	Mid Height
	4/12/63	Epoxy-Cemented	0.60	Near Top
	4/12/63	Epoxy-Cemented	0.60	Near Top
1-3	6/21/62	Flange-Clamped	0.50	Near Top
	6/21/62	Flange-Clamped	0.52	Near Top
	7/ 7/62	Flange-Clamped	0.53	Near Top
	7/ 7/62	Flange-Clamped	0.55	Near Top
	7/11/62	Epoxy-Cemented	0.51	Near Top
3-1	7/15/62	Epoxy-Cemented	0.61	Near Top
	7/16/62	Epoxy-Cemented	0.61	Near Top
4-1	7/14/62	Epoxy-Cemented	0.79	Near Top
	4/5/63*	Epoxy-Cemented	0.86	Near Top
	4/7/63*	Epoxy-Cemented	0.90	Near Top
5-1	7/27/62	Epoxy-Cemented	0.99	Near Top
	3/11/63	Epoxy-Cemented	1.11	New Spot Near Top
	3/12/63	Epoxy-Cemented	1.11	New Spot Near Top
	3/13/63	Epoxy-Cemented	1.11	New Spot Near Top
5-3	7/31/62	Epoxy-Cemented	0.77	Near Top
	9/ 7/62	Epoxy-Cemented	0.84	Near Top
	5/ 7/62	Epoxy-Cemented	0.77	Near Top
6-1	6/23/63	Epoxy-Cemented	1.00	Near Top
	6/24/63	Epoxy-Cemented	1.01	Near Top
6-2	6/21/63	Epoxy-Cemented	1.00	Near Top
	6/25/63	Epoxy-Cemented	1.00	Near Top
6-3	6/20/63	Epoxy-Cemented	1.02	Near Top
	6/26/63	Epoxy-Cemented	1.02	New Spot Near Top
	6/27/63	Epoxy-Cemented	1.01	Same as 6/20/63
101-1	6/28/63	Epoxy-Cemented	0.85	Near Top
	7/ 4/63	Epoxy-Cemented	0.90	Near Top
101-2	7/ 2/63	Epoxy-Cemented	0.71	At Top
	7/ 5/63	Epoxy-Cemented	0.78	At Top
	7/ 7/63	Epoxy-Cemented	0.78	At Top

*To check the effect of holes for hanging weights.

Table 4.3. Shell Thickness Data

Shell	t_{min} (inches)	t_{max} (inches)	$\frac{t_{max} - t_{min}}{t_{min}}$, %	t_{ave} (inches)	$t_{at\ buckle}$ (inches)
1-1	0.0240	0.0260	8	0.0253	0.0255
1-2	0.0235	0.0260	11	0.0253	0.0250
1-3	0.0235	0.0265	13	0.0252	0.0250
2-1	0.0225	0.0275	22	0.0252	0.0265
2-2	0.0220	0.0280	27	0.0254	0.0270
2-3	0.0225	0.0285	27	0.0261	0.0270
3-1	0.0255	0.0285	12	0.0257	0.0255
3-2	0.0230	0.0255	11	0.0241	0.0230
3-3	0.0230	0.0255	11	0.0242	0.0235
3-4	0.0240	0.0270	11	0.0251	0.0240
3-5	0.0215	0.0250	12	0.0239	0.0220
4-1	0.0260	0.0310	19	0.0283	0.0300
4-2	0.0250	0.0305	22	0.0284	0.0300
4-3	0.0260	0.0310	19	0.0289	0.0290
5-1	0.0275	0.0310	13	0.0298	0.0305
5-2	0.0250	0.0300	20	0.0281	0.0280
5-3	0.0250	0.0300	20	0.0277	0.0275
6-1	0.0280	0.0325	16	0.0304	0.0320
6-2	0.0280	0.0320	14	0.0303	0.0315
6-3	0.0290	0.0335	16	0.0310	0.0320
101-1	0.0600	0.0630	5	0.0616	0.0615
101-2	0.0600	0.0635	6	0.0617	0.0600
102-1	0.0585	0.0640	9	0.0621	0.0630
102-2	0.0605	0.0650	7	0.0629	0.0645

average thickness plot on a slope of 2.9/1. Thus, p_{cr}/E is more nearly proportional to the 2.9 power of t/R than to the second power as predicted by theory. Some people would try to explain this by saying that imperfections are what cause experimental buckling pressures to be lower than theoretical pressures, and then surmising that the imperfections present were less significant the thicker the shell.

When the test data of all investigators of cylindrical tubes and spherical caps are plotted, there is a definite trend of increasing discrepancy between classical theory and experiment the larger the R/t ratio. In 1934, Donnell[9] had postulated that the difference between experiments and classical theory was due to initial imperfections. He later extended this concept by using a relationship between the initial imperfections and the buckle geometry. This combined relationship, which he named the "unevenness" factor U_0, allowed the derivation of a relationship between $p_{cr}/p_{classical}$ and U_0R/t. Donnell found it necessary to determine U_0 in order to fit his theory to any test result, but his theory did lead to larger deviations from classical linear theory for larger and larger R/t ratios. Gerard and Becker[16] extended the "unevenness" concept to spherical shells by empirically fitting a curve to the test results of Kaplan and Fung.[21] They propose that

$$U_0 = A_0 S$$

where

U_0 = "unevenness" factor

A_0 = ratio of the amplitude of the equivalent imperfection sine wave to the shell thickness. This theoretically includes not only geometric imperfections but also residual stresses, material anisotropy, and loading eccentricities.

S = "sensitivity" factor which measures the sensitivity of the imperfection in terms of the buckle wavelengths.

They compute U_0 for each of Kaplan and Fung's test shells. Then the empirical curve shown in Figure 4.24 can be drawn. On the assumption that the "unevenness" concept is valid, the present test results are compared to the "unevenness" theory in Figure 4.25. It is seen that any test result can be "explained" by choosing the proper value for U_0. On the other hand, no single value of U_0 fits the trend of the experimental result. It must be pointed out that the first tests conducted were for the thinner shells, and then progressively thicker shells were tested. Perhaps the experimental technique did improve throughout the course of the program, but it is felt that this technique did not improve by the factor of 2 (that is, $0.000040 < U_0 < 0.000075$), which would be required to explain the trend line on Figure 4.25.

In a design study, the ultimate goal is to extrapolate the model result to the prototype. If the "unevenness" concept is correct, it would be proper to have the same unevenness in the model as would be encountered in the prototype. Due to the nature of the "unevenness" factor, it would be impractical to predict its prototype magnitude even for the spherical cap, let alone some complex form. Any amount of error involved in going from a model result to a prototype would be indeterminate and dependent upon factors beyond the control of the engineer. At the present time, it would have to be incorporated into the safety factor.

The test results are also shown in Figures 1.7b and 1.9 where they can be compared to the results of other investigators. Again,

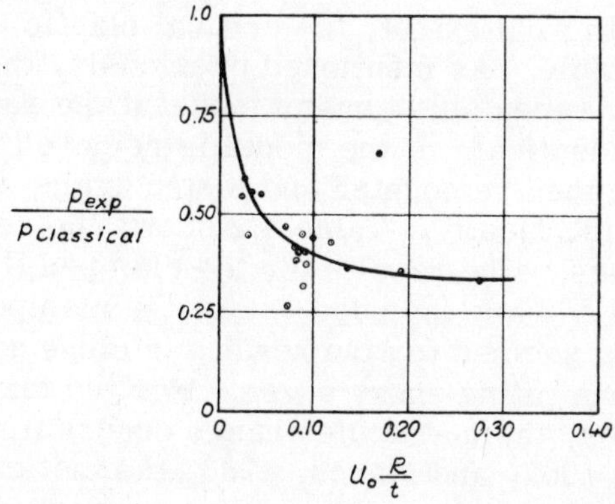

Figure 4.24. Empirical curve fitted through spherical shell test data of Kaplan and Fung to obtain relation between critical pressure and "unevenness" factor.[38]

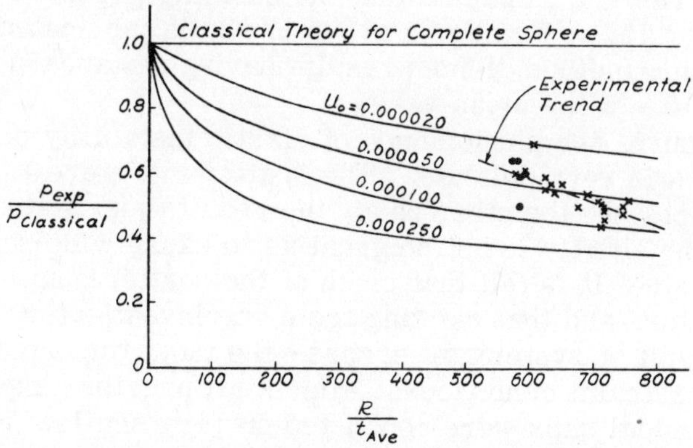

Figure 4.25. Buckling pressures versus R/t as derived from Figure 4.24. Present test results compared.

it is noted that for the thicker shells the results more nearly approach the theoretical value.

Now the primary reason for making these tests was to clarify whether, and to what extent, the critical elastic buckling pressure is repeatable. As mentioned previously, the pressure is almost exactly repeatable for any model if the support condition is absolutely identical. If the model is removed from the test equipment and then remounted and tested again, the pressures do not repeat identically. Table 4.2 shows the results of these successive tests. The results for the clamped flange tests on shells 1-1 and 1-3 are included, since the clamped edge for the series 1 shells seemed to give results in close agreement with the tests incorporating epoxy edges. For the texts with epoxy-cemented edges, the maximum change occurs for shell 5-1 where, between July, 1962, and March, 1963, the critical pressure increased 13%.

Figure 4.26 deals with repeatability from one shell to another and from one mold to another. For the twenty model shells the mean value of $p_{exp}/p_{classical}$ equaled 0.54 while the standard deviation was 0.058. If it had not been for shell 5-1, the dispersion would have been much smaller.

In sum, it can be concluded that, for any one shell, the buckling phenomena are repeatable to within 1% if the boundary conditions are absolutely identical. If a shell is removed from the testing rig and then remounted and tested again, the results listed in Table 4.2 indicate that the buckling pressure may vary up to 10 or 15%. Finally, a group of shells fabricated and tested by the same method yielded results having a standard deviation of about 10% of the mean value.

To be sure, the phenomenon of elastic instability of thin-shell structures is very delicate. The scatter in Figure 1.7b attests to this fact. On the other hand, the results just stated indicate that the sensitivity is not so great as to cause wide scatter in repeated tests. It is felt that much of the scatter obtained by some investigators and that existing from one investigator to another is the result of systematic errors—the most important being in the end restraint conditions. Almost all previous experiments with spherical caps were conducted on very shallow shells. In such shallow shells, edge disturbances penetrate throughout the entire shell. Even though the present shells are not shallow, Figure 4.27 shows what serious changes can be effected by a change in edge restraint conditions. Obviously, flange-clamping introduced a significant edge bending which caused a premature instability to arise at the shell edge. All of the series 2, 3, and 4 flange-clamped shells buckled at the edge. The edge effect was due to a small fillet that occurred at the shell-flange junction of the models formed over green, relatively cold, plaster molds. The wooden mold No. 1 did not cause such rapid cooling of the plastic, and as a result, the transition from flange to shell was sharper.

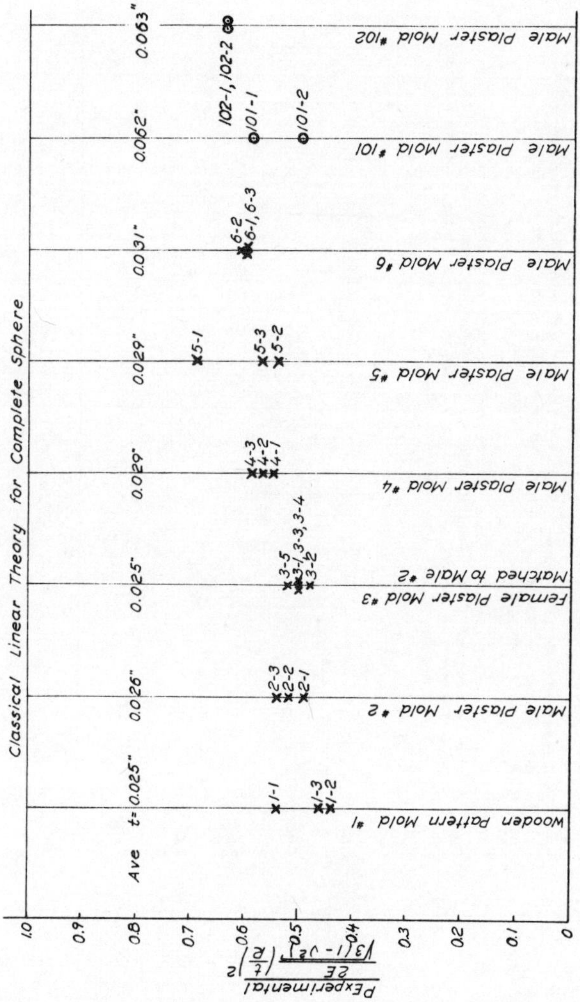

Figure 4.26. Repeatability of buckling pressures in spherical shells.

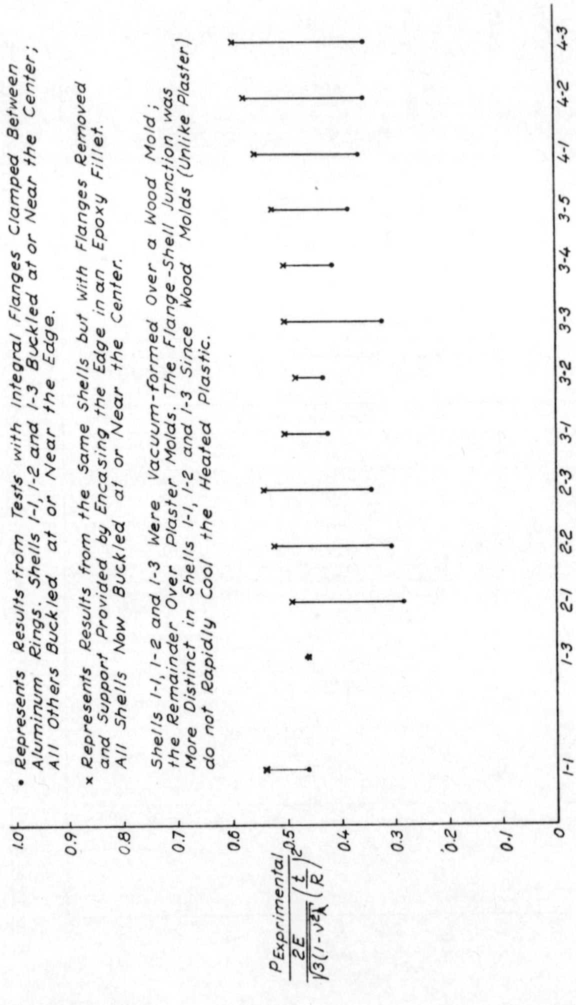

Figure 4.27. Flange-clamped versus epoxy-cemented edges.

One final observation concerns the question of whether thickness variation has a significant influence on the critical pressure. Table 4.3 summarizes thickness data. Series 1, 2, 4, 5, and 6 shells, being formed on a male mold, had maximum thicknesses near the top. On the other hand, the series 3 shells were formed into a female mold, and maximum thicknesses occurred near the edges. Restricting one's attention to the behavior of the matched mold series 2 and 3, some important points should be noted.

The buckle position always occurs near the top, in series 2 at a place having a thickness greater than the average thickness and in series 3 at a place with a less than average thickness. It is not clear why some of the series 2 shells (and others which were thinnest at the edges) did not buckle at the edge. The first explanation is that the epoxy edge condition prevented the buckling. On the other hand, later vertical weight-loading tests led to edge buckling in all cases. For the vertical weight-loading condition, the meridional membrane stress resultant at the shell edge is 7% greater than that at the top; however, the edge membrane stresses in the radially loaded series 2 shells are about 15% greater than those at the top.

In spite of the lack of understanding about the buckle position, the magnitude and distribution of thickness variation in the series 2 and 3 shells did not have an effect upon the correlation of buckling pressures and average thickness. The magnitude of the total thickness variation in the series 2 shells is about twice that in the series 3 shells and the nature of the variations are reversed, that is, where the series 2 shells are the thickest, the series 3 shells are the thinnest, and vice versa. These results indicate that thickness variations are not the controlling factor in thin-shell buckling models, and this fact is of great importance to the model fabricator.

4.3 Air Pressure versus Discrete Weight Loading

For the laboratory test, fluid pressure loading is relatively easy to mobilize and measure. Unfortunately, a normal pressure loading is most unusual in practical thin-shell roof constructions. Of course, if the shell is flat enough, then one can perhaps approximate a gravity loading by a normally directed one. In another vein, the mobilization of fluid pressure loadings on structural models demands either that the boundary be continuously supported or that the shell shape be compatible with the use of an air bag system. Therefore, the model requirements often demand a departure from the pressure concept. Even though the amount of, and tedium of, the work involved becomes larger, the use of hanging weights has several advantages. Hanging weights do simulate gravity load, they do not demand any particular type of boundary support, they permit unsymmetrical and partial loadings to be applied, and they will not restrain the model. On the other hand, hanging weights imply a discrete loading rather than the desired continuous system and require some kind of attachment mechanism which could influence the structural behavior.

Four of the twenty 18-inch radius air pressure models were loaded by hanging weights in order that the buckling behavior under air pressure and discrete vertical weight loading could be compared. Consideration was also given to conducting an intermediate test with discrete radial loads. Difficulties in devising and constructing an appropriate radial system led to the abandonment of this step.

4.3.1 Weight-Loading Testing Technique. The primary difficulties in designing an acceptable support system for a hanging weight test arise from the fact that there must be a load on – load off device. The time and spatial discontinuities inherent in the addition of load increments must not be allowed to influence the model. Further, it is greatly to be desired that some provision be made to "catch" a buckled model before it is destroyed.

To meet these requirements, the wooden mold No. 1, which was used to form the series 1 models, was adapted to support the models before loading and to "catch" them after buckling. Figure 4.28 shows the mold, with holes at a 1-inch surface grid, as it fit into the testing structure. Raising and lowering of the mold was accomplished through two oil-hydraulic jacks. When the mold was in the raised position, the model would bear directly against the mold and be supported by it. When in this position, load increments could be added or taken away. To load the model, the jacks could be lowered at the rate of about 1/2 inch per minute. The load would be gradually and uniformly transferred to the model. If the model buckled under the load, it would simply follow the mold down. If buckling was observed, the mold was raised again until the load was removed from the shell. The 3/4-inch plywood, to which the model edges were epoxy-cemented, merely separated from and beared upon the supporting structure when the model load was off and on respectively. Figure 4.29 shows the entire system.

4.3.2 Results and Conclusions. The information obtained from these tests concerned three things, namely: (1) the effect of load grid spacing, (2) the effect of boundary condition changes, and (3) the manner in which the buckling behavior compared to the air pressure tests. Weight increments of 0.525, 0.264, 0.136, and 0.026 pound were used. An account of the tests and data from them are contained in Appendix C. Buckle positions are shown on the shell information sheets in Appendix B.

Load application corresponding to three different grid spacings was applied. First, 241 holes, 0.043 inch in diameter, were

Figure 4.28. Supporting and catching mold for weight tests.

Figure 4.29. Testing equipment for weight tests.

drilled through the models on a 1-inch surface grid. Strings were passed through the holes and connected to one of the three-legged loading pads shown in Figure 4.30. The rubber legs are sufficiently flexible so that the loading pad does not stiffen the model. Loading shell 1-1 through the 241 load pads, as shown

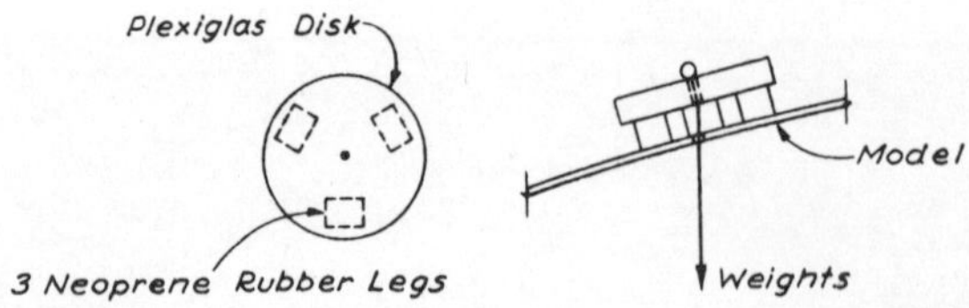

Figure 4.30. Detail of loading pad.

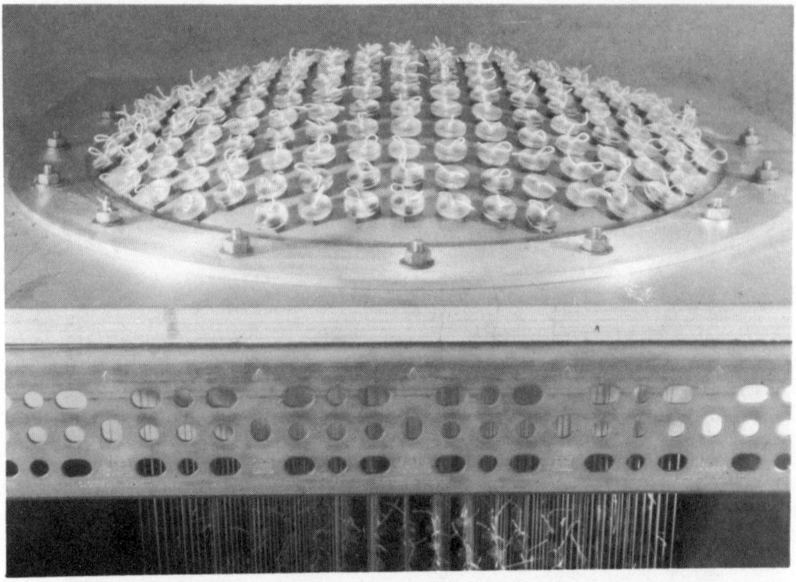

Figure 4.31. Loading pads at 1-inch grid spacing.

in Figure 4.31, led to a critical load of 0.536 pound per string. The loading pads were then removed, and with the string knots bearing directly on the shell as in Figure 4.32, the same critical load was obtained. Finally, when the load was applied over a 2-inch grid, the deflected shape prior to buckling was so distorted that the shell no longer corresponded to a spherical dome. These deformations are shown in Figure 4.33. Surprisingly enough, this load pattern sustained 2.247 pounds for each of the 61 strings—or more per unit area than the two previous load conditions. It is felt that the number of load points must be sufficient to insure that severe local dimpling does not occur prior to buckling. For the present R/t ratio (≈ 650), both the direct 1-inch grid and the 1-inch grid with loading pads are satisfactory. Larger grid spacings could be used for smaller R/t ratios.

For the air pressure tests, a significant buckling pressure change was noted when the technique for supporting the edge was altered from flange clamping to epoxy encasement. On the other

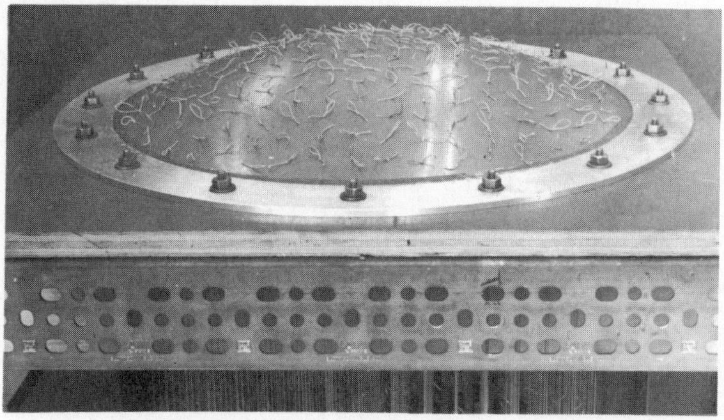

Figure 4.32. 1-inch loading grid without pads.

Figure 4.33. 2-inch loading grid without pads.

hand, with epoxy encasement, variations in the size of the epoxy fillet had no effect on critical pressure. Since all buckle positions were near the top of the shell in the air pressure tests, there was no assurance that variations in the size of the epoxy fillet would not be crucial when the buckle position was at the edge—as was the case for the hanging weight tests. Results with shell 1-1 confirmed the doubt. The "normal" epoxy fillet which was used for all air pressure tests (except those where deliberate changes were made to see if the results would change) is shown in Figure 4.18. Using a 1-inch grid spacing with loading pads, and a slightly smaller than "normal" fillet, shell 1-1 buckled at the edge under a load of 0.400 pound per string or psi. The top or exposed portion of the fillet was then carefully stripped away, and then, a retest yielded a critical pressure of only 0.264

psi. The shell was removed and recemented with a slightly larger than "normal" fillet. The buckle position remained the same, but now the shell would hold 0.536 psi. Thus, the edge condition is extremely critical. Like any two-part epoxy, the stiffness of the cured resin depends upon mix proportions and environmental conditions during cure; consequently, epoxy hardness variations could also affect the result.

Table 4.4. Summary of Test Data for Domes Loaded by Hanging Weights

Shell	1-1	1-3	3-1	4-1
Radial Critical Pressure (psi)	0.570	0.522	0.610	0.792
$\frac{\text{Radial Critical Pressure}}{1.07}$	0.531	0.487	0.570	0.740
Vertical Critical Pressure (psi)	0.536	0.588	0.606	0.663
Thickness at Rad. Buckle (top)	0.0255	0.0250	0.0255	0.0300
Thickness at Vert. Buckle (edge)	0.0240	0.0240	0.0263	0.0270
Average Thickness (in.)	0.0253	0.0252	0.0267	0.0289

Table 4.4 summarizes the test data. Comparison of the membrane stress state for radial and vertical pressure (Figure 2.2) shows that, unlike the radial case, vertical pressure leads to a varying meridional stress. At the apex of the dome, the meridional stress for radial and vertical pressure is equal, but the vertical pressure stress becomes relatively larger as one goes down the sphere. For the particular opening angle of the tested shells, the increase is 7% at the shell edge. On the assumption that the meridional stress is much more significant than the circumferential stress, the air pressure critical loads are "normalized" by the factor 1.07. Comparing the vertical weight and air pressure results on this basis, the weight tests gave pressures 21, 6, and 1% larger and 12% smaller.

All four of the tested shells buckled, initially, at the thinnest spot around the shell edge. As the "catching" mold sank lower and lower, separate buckles would occur over the entire shell. Figure 4.34 shows the buckle as the shell follows the "catching" mold.

According to the behavior of shells 1-3 and 3-1, the degree of edge restraint has a much more significant effect on buckling pressures than do thickness variations. Shell 1-3, which is

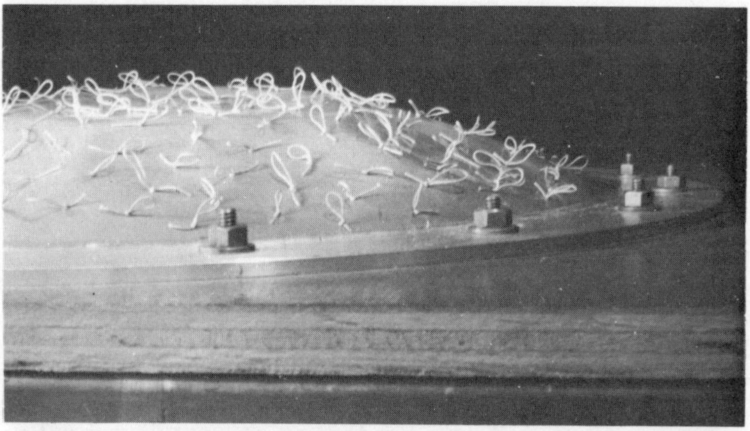

Figure 4.34. Buckled shape for vertical weight loading.

Figure 4.35. Air pressure test after filling string holds.

thinnest at the edge, buckled at the edge under a load 21% greater than the radial critical pressure. Shell 3-1, formed into a female mold, was thicker at the edge than at the top, yet showed only a 6% increase over the radial pressure. Likewise, shells 1-3 and 4-1 have similar thickness variations, yet shell 1-3 showed a 21% increase while shell 4-1 showed a 12% decrease.

In order to see if the drilled holes had any effect in reducing stiffness, shell 4-1 was retested under air pressure. The holes were filled with plasticene clay, as shown in Figure 4.35. The resulting air critical pressure was even higher than it had been before, increasing from 0.792 psi up to 0.862 psi. A second retest gave 0.897 psi. Thus, if the holes cause a significant stiffness reduction, this reduction was masked by unknown factors.

Two important conclusions can be drawn from the hanging weight tests. First, it is apparent that buckle position is sensitive to load direction. How, and to what extent, the buckle position affects critical buckling pressure is unknown, but it is recommended that prototype gravity loads be reproduced as gravity loads in a model unless the shell is very flat. In this way the model behavior will at least be occurring in the right place, and this behavior may alert the model investigator to possible sources of systematic error.

Second, the scatter in the results of the hanging weight tests is far greater than that experienced for the air pressure tests. This scatter is surely caused by the variation in edge restraint. An apparently defective can of epoxy cement was used for the weight tests, and relatively large variations in epoxy hardness resulted. This situation, coupled with the fact that the shell wanted to buckle right at the edge, was serious. In the weight tests, then, the edge condition is even more critical than in the air pressure tests—and in the air pressure tests it will be remembered (Figure 4.27) that pressure changes of almost 100% were caused by changes in edge condition. The conclusion is that any buckling model must reproduce as nearly as possible the prototype boundary condition; and if the prototype condition is uncertain, then provision should be made for alternate model tests in which the possible prototype conditions are bracketed.

4.4 36-Inch Radius Domes Subjected to Air Pressure

Four shell models from two molds were fabricated and tested. These models were completely similar to the 18-inch domes, but double in size. The experimental procedure and results obtained are included in this section. For identification purposes shell 101-1 indicates that the model was the first one pressed on mold No. 101.

4.4.1 Fabrication and Testing Procedure. Two plaster molds were constructed using Ultracal 30 gypsum plaster. The manufacturing process was the same as that used for the 18-inch domes. An accurately machined screed was rotated about a 1/2-inch Plexiglas base as shown in Figure 4.36. A semirigid plastic foam filler material was used to lighten the first mold, but was not included in the second mold. After screeding, the mold would look as shown in Figure 4.37. A finished mold is shown in Figure 4.38. Vacuum-forming was accomplished on a commercial 42×72 inch machine.

The testing procedure was exactly the same as that used in the 18-inch radius domes, except that all tests incorporated epoxy cement edge restraint. The test equipment is shown in Figure 4.39. It perhaps should be noted that the auxiliary air chamber tank which is shown in Figure 2.16 was used for both the 18- and

Figure 4.36. Screed for 36-inch radius mold.

Figure 4.37. Screeded 36-inch radius mold.

Figure 4.38. Finished 36-inch radius mold.

Figure 4.39. Air pressure test equipment for 36-inch radius domes.

36-inch radius tests, but its presence or absence had no effect upon any buckling pressures.

4.4.2 Data. A summary of the factors attendant to each of the four model shells and the data regarding the buckling capacity of the shells is presented in Appendix D.

4.4.3 Results and Conclusion. The results from the four model shells are included with the 18-inch radius model results in Tables 4.1 to 4.3 and Figures 4.23, 4.25, and 4.26. The behavior of the large models was completely analogous to that of the small models.

A comparison of the buckling pressures of the two series 101 models and the two series 102 models does lend credence to the imperfection theory. Mold 101 cracked during the forming of shell 101-1, and these cracks became even worse when shell 101-2 was pressed. The resulting models had crack marks in several places. The marks started at the shell edge and worked their way approximately halfway to the top. In model 101-2 the change in elevation from one side of a mark, to the other, was about 0.005 inch. Under ordinary circumstances, these models would have been rejected; however, because of the monetary investment in them, they were tested anyhow. The fact that the buckle position on both of the models occurred at the top, unmarked portion was the only reason why the test results were not rejected.

In spite of the mold cracking problem, it can still be concluded that changing the geometric scale by a factor of 2 did not significantly affect the model behavior. The time and cost associated with a model study is significantly affected by the geometric scale, and hence, it is important to know whether the chosen model-prototype scale ratio will influence the result.

4.5 Summary of Experimental Conclusions

1. Small-scale plastic models offer a reliable means of obtaining information concerning the elastic stability of thin-shell structures.
2. At the strain levels encountered in thin-shell buckling problems, the properties of Boltaron 6200 PVC can be taken as elastic. Further, these elastic properties will not vary widely throughout a vacuum-formed shell model and can be determined within 5%.
3. Creep will not be a controlling factor at the strain levels encountered in buckling studies with Boltaron 6200 PVC shell models.
4. Average shell thickness should be taken as the controlling thickness quantity in a shell buckling study. In this light, minor thickness variations (that is, ± 10% or less) can be tolerated since they seem to have little influence on critical pressures.
5. Buckle position is influenced by factors other than thickness variation. Thus all radially loaded models buckled near the top, while all weight-loaded models buckled at the shell edge. On the other hand, all edge buckles occurred at the thinnest spot around the edge.
6. Buckling pressures were proportional to $(t/R)^{2.9}$ instead of $(t/R)^2$ as predicted by the classical theory. This difference may be due to "unevenness" in the models. To account for such "unevenness" in a design program would be impractical if not impossible. The experimental stress analyst should manufacture his model as carefully as possible, but beyond that, he cannot worry that the subsequent buckling behavior of his model (and the as yet imaginary prototype) is being controlled by imperfections which he simply cannot control. Such uncertainties must presently be accounted for in the safety factor.
7. Soosaar[40] has shown that, for shell stress-distribution studies, the replacement of a continuous loading by a grid of discrete loads will lead to very large systematic errors. Experience with ultimate strength studies of microconcrete shell models again indicates that discrete loadings may lead to erroneous results (for example, punching shear failure). On the other hand, the substitution of a discrete load pattern does not affect the buckling pressure if the grid spacing is small enough. For all practical cases of R/t ratios it should never be necessary to use less than a 1-inch grid spacing.
8. The buckling behavior of any thin-shell model may be very sensitive to changes in boundary condition. Thus, any buckling model must reproduce as nearly as possible the

prototype boundary condition; and, if the prototype condition is uncertain, then provision should be made for alternate model tests in which the possible prototype conditions are bracketed.

9. The buckling behavior of the shells was not affected by a 100% change in the geometric scale.

Chapter 5

A SUMMATION

Two examples of the use of small-scale design models were noted in Chapter 1. The problem of instability, or buckling, was predominant in each of these studies. Other examples could be cited wherein the concern lay not with buckling but rather with questions concerning stress-distribution, or ultimate strength, or both, or all of them. Most commonly, economic factors prohibit the testing of more than one or two models; and so, the engineer does not have the chance to check his results by repeated testing. There are many references in the literature to model tests on thin-shell structures wherein the data obtained could not be interpreted satisfactorily. These difficulties arose because of the incorporation of systematic errors at some stage of the experimental program.

One of the ironies of a mathematical model is that the distortions of truth which are introduced by improper assumptions or inaccurate models remain secluded in the mist of one's imagination. With a physical model, the results of such improprieties are dramatically exposed. The philosophical effect of this distinction between the mathematical and experimental methods of design is, on the one hand, a detriment to the experimental approach, but at the same time, can be one of its salient features. Thus, meaningful model results are just as dramatic in a positive way as erratic ones are in a discouraging way.

Successful experimental shell buckling studies have been made, and this monograph supports the proposition that small-scale plastic models should yield reliable results in this regard. The fact that PVC material properties can be closely determined, that creep is not a serious problem, that minor thickness variations are not significant, that there is no scale effect, and that discrete load systems can be substituted for continuous ones, lead one to have faith in a model result. The fact that there is very little or no instrumentation involved in a buckling model study eliminates the myriad of potential error sources associated with any instrumentation system. The sensitivity of thin-shell buckling to initial imperfection demands that the model be carefully fabricated, but surely a physical model is better able to deduce such sensitivity than our present mathematical models. On the other hand, it should be possible to improve upon our present means of having to account for imperfection effects by a safety factor. Such research should incorporate the use of photogrammetric means for establishing the surface geometry of the test shells. Finally, it must be noted that a thin-shell buckling study may be extremely sensitive to the means of providing boundary support. The experimental designer should make every effort to reproduce, exactly, the prototype conditions.

It is hoped that an increased use of physical models will be seen in the coming years. They have not received the attention which they deserve, and too often, their use has been maligned by results obtained by incompletely informed engineers.

APPENDIX A

Miscellaneous Data Regarding Modulus Tests, Poisson's Ratio Tests, Edge Restraint Tests, and Residual Stresses

Date:	July 6, 1963		
Temperature:	79°F		
Relative Humidity:	Unknown		
Specimen:	Test Sample 6-1-A	Width	= 0.699"
		Thickness	= 0.0315"
	Selected such that $E_{bending}$ = 475,000 psi		
Active and Dummy Gages:	1 vertical gage attached to each face		
	Baldwin-Lima-Hamilton	FA - 25-12 S-6	
		G.F. = 2.03 ± 1%	
		R = 120.0 ± 0.2 Ω	
Cement:	Baldwin-Lima-Hamilton	EP 150	
Strain Indicator:	Baldwin-Lima-Hamilton, Type N, Serial No. 651761		

Load	Stress	One Gage in Arm 1 of a Half Wheatstone Bridge		Opposite Gage in Arm 1 of a Half Wheatstone Bridge		Gages in Arms 1&3 of a Full Wheatstone Bridge
		Run #1	Run #2	Run #3	Run #4	
zero	0	0- 8-1285	0- 8-1295	0- 8-1080	0- 8-1085	0- 8-1005
5 lbs.	241 psi	0- 8-1650	0- 8-1660	0- 8-1435	0- 8-1435	0- 8-1730
10	483	0-10-0015	0-10-0025	0- 8-1795	0- 8-1790	0-10-0450
15	724	0-10-0380	0-10-0390	0-10-0150	0-10-0150	0-10-1190
20	966	0-10-0750	0-10-0760	0-10-0515	0-10-0515	0-10-1925
15		0-10-0390	0-10-0400	0-10-0160		0-10-1190
10		0-10-0030	0-10-0030	0- 8-1810		0-10-0455
5		0- 8-1670	0- 8-1670	0- 8-1450		0- 8-1720
zero		0- 8-1305	0- 8-1305	0- 8-1095	0- 8-1095	0- 8-0995

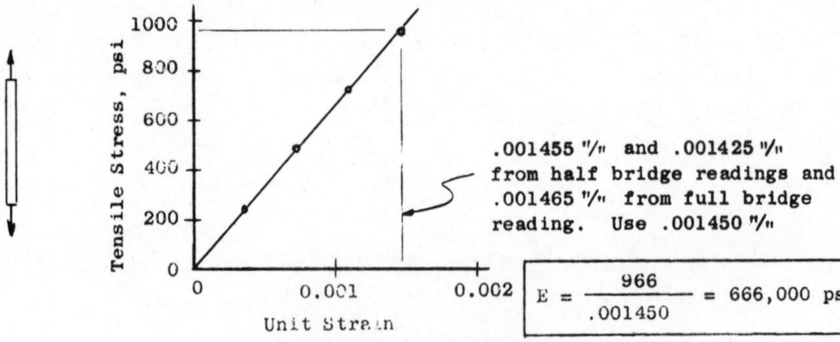

$$E = \frac{966}{.001450} = 666,000 \text{ psi}$$

Plate A-1 SR-4 Gage Tests for Tensile Modulus of Boltaron 6200 PVC

Date:	**July 6, 1963**
Temperature:	**79°F**
Relative Humidity:	**Unknown**
Specimen:	**An available 1/16" piece** Width 0.691" Thickness = 0.0647"
	Selected such that $E_{bending}$ **= 475,000 psi**
	1 vertical **gage attached to each face**
Active and Dummy Gages:	Baldwin-Lima-Hamilton FA - 25-12 S-6
	G.F. = 2.03 ± 1%
	R = 120.0 ± 0.2 Ω
Cement:	Baldwin-Lima-Hamilton EP 150
Strain Indicator:	Baldwin-Lima-Hamilton, Type N, Serial No. 651761

Load	Stress	One Gage in Arm 1 of a Half Wheatstone Bridge		Opposite Gage in Arm 1 of a Half Wheatstone Bridge		Gages in Arms 1&3 of a full Wheatstone Bridge
		Run #1	Run #2	Run #3	Run #4	
zero	0	0-10-1490	0-10-1495	0-12-0975	0-12-0970	0-12-1085
5 lbs.	112 psi	0-10-1680		0-12-1155	0-12-1150	0-12-1460
10	224	0-10-1875		0-12-1335	0-12-1330	0-12-1840
15	335	0-12-0065		0-12-1520	0-12 1515	0-14-0220
20	447	0-12-0255	0-12-0260	0-12-1705	0-12-1700	0-14-0605
15		0-12-0070		0-12-1520	0-12-1515	0-14-0220
10		0-10-1880		0-12-1335	0-12-1330	0-12-1840
5		0-10-1685		0-12-1155	0-12-1150	0-12-1470
zero		0-10-1500	0-10-1500	0-12-0970	0-12-0970	0-12-1090

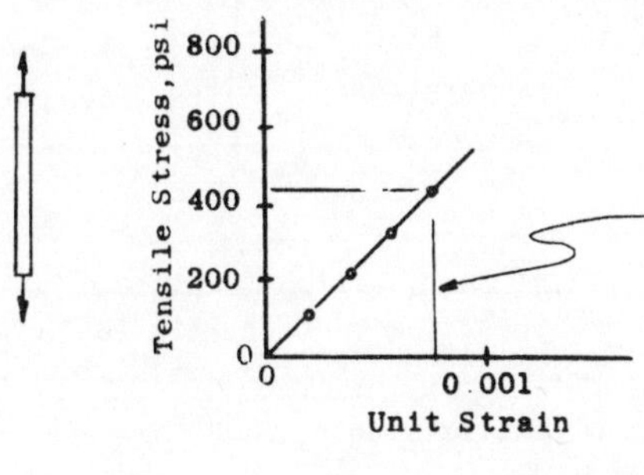

.000760 "/" and .000735 "/" from half bridge readings and .000760 "/" from full bridge reading. Use .000750 "/"

$$E = \frac{447}{.000750} = 596,000 \text{ psi}$$

Plate A-2 **SR-4 Gage Tests for Tensile Modulus of Boltaron 6200 PVC**

Date	April 1, 1962
Temperature	72°F
Relative Humidity	50%
Specimens	Six specimens (3 in each direction) cut from the edges of a vacuum formed part. Specimens machined to ASTM D638 - 60T specifications Width at grips = 0.751 ± 0.001" Width at gage = 0.501 + 0.002" Thickness = 0.065 ∓ 0.001"
Apparatus	Instron Testing Machine
Gage	The extensometer originally used was in error. The results presented here are corrected on the basis of subsequen tests using the LVDT extensometer shown in Figure 4.2.

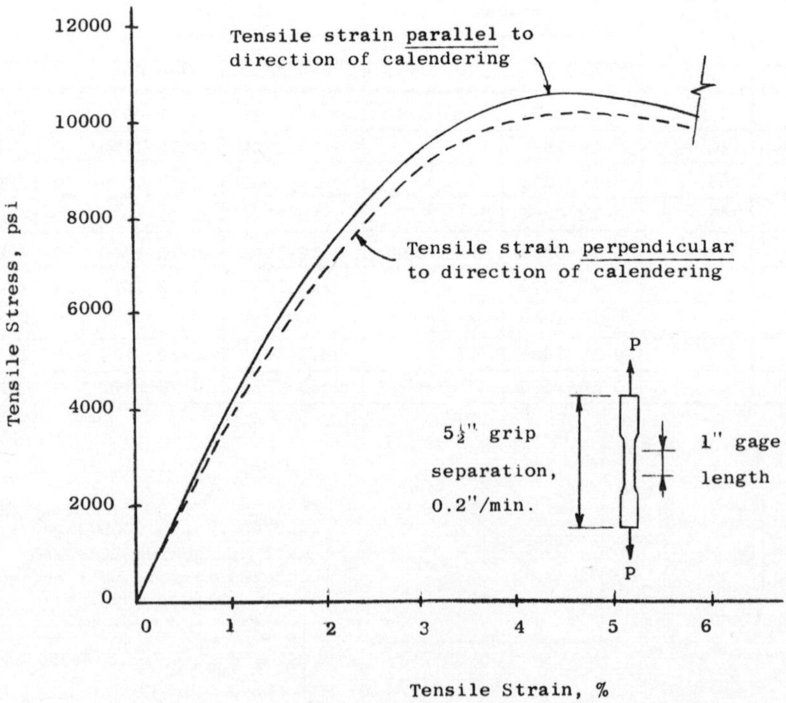

Plate A-3 Stress-Strain Curve from Mechanical Gage on Boltaron 6200 PVC

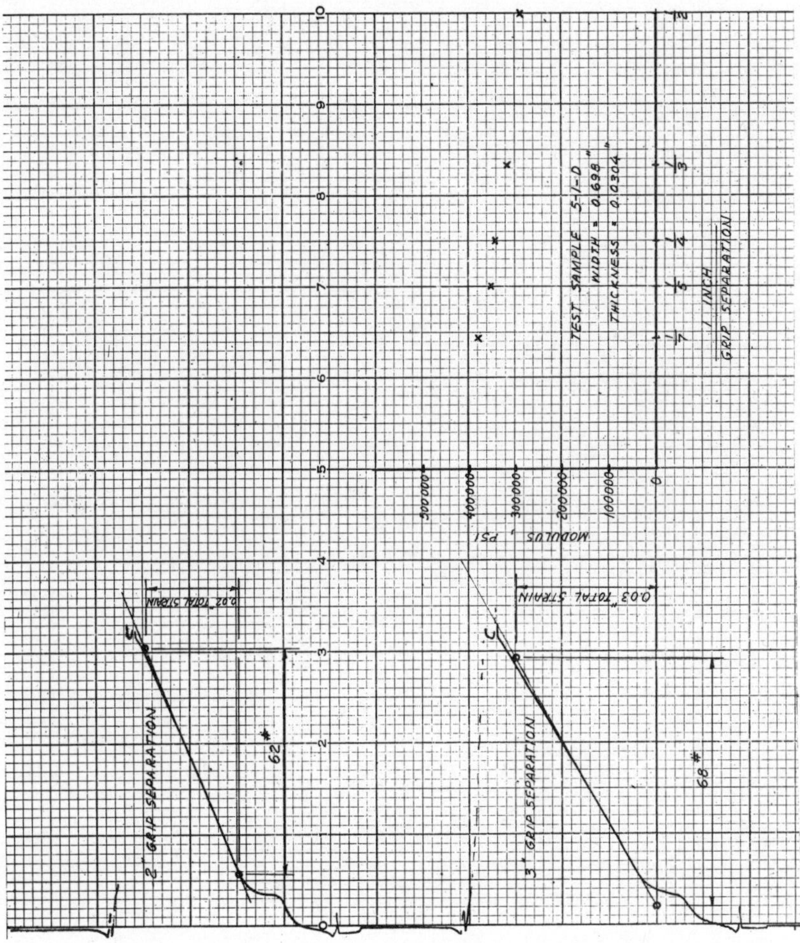

Plate A-4 Typical Data Sheet for Tensile Modulus of Boltaron 6200 PVC from Instron Testing Machine

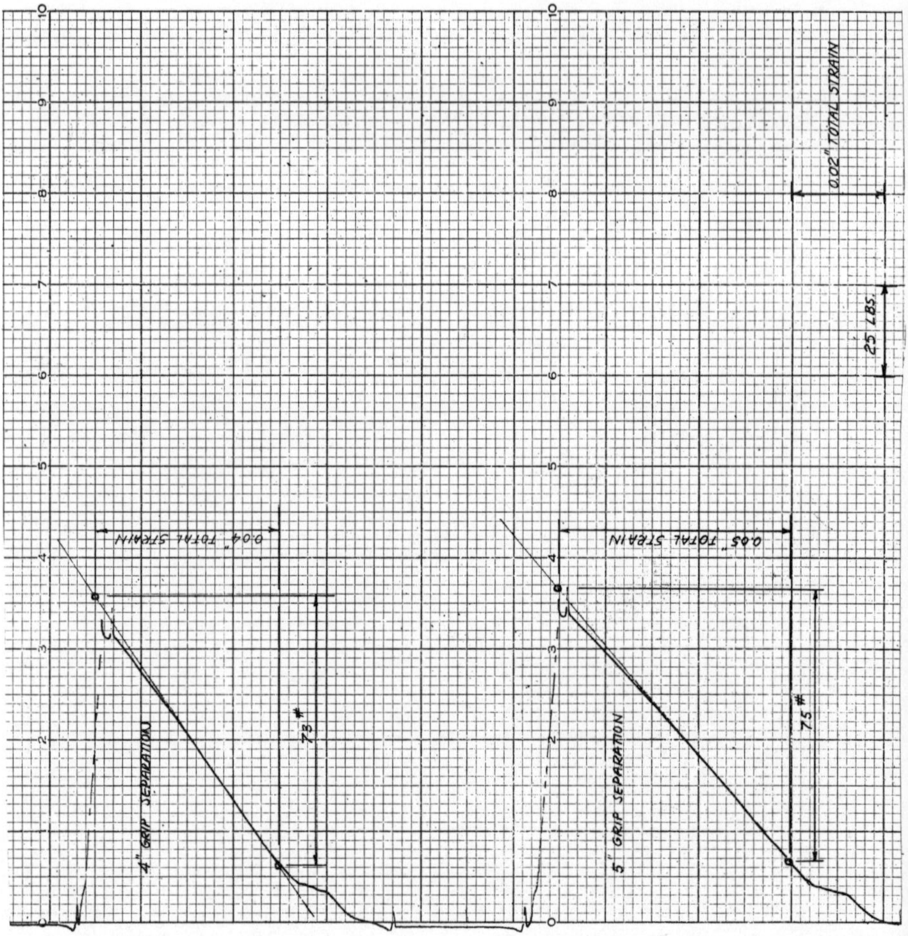

Plate A-4 (Continued)

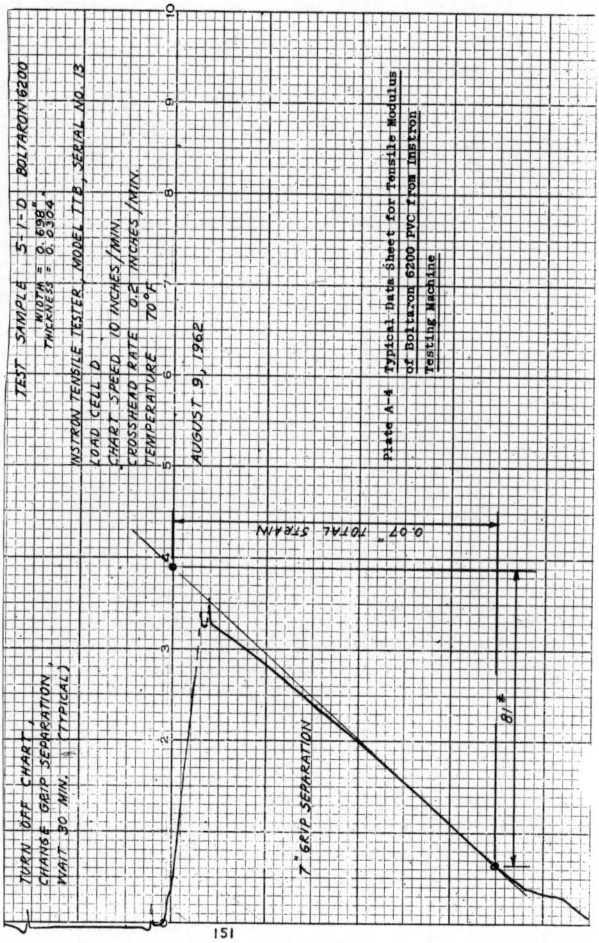

Plate A-4 (Continued)

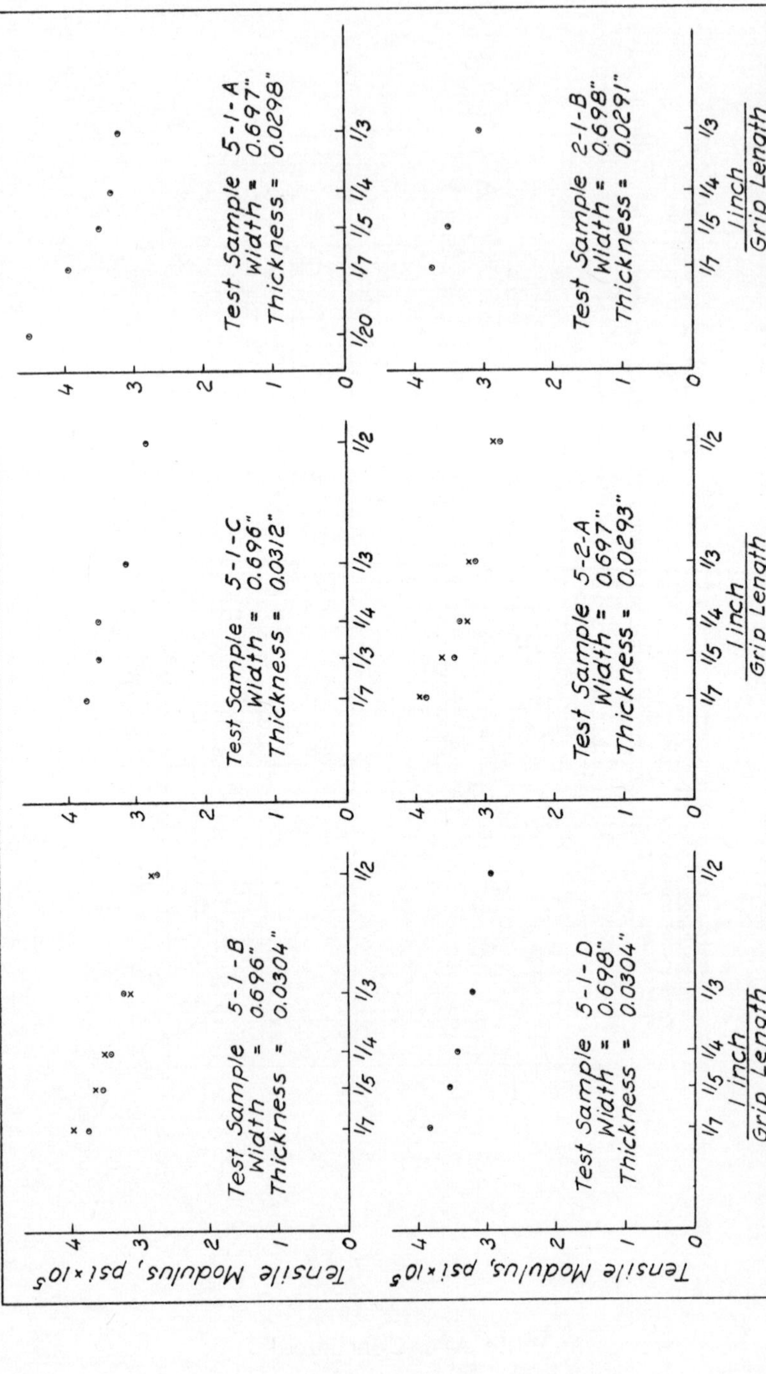

Plate A-5 Effect of Grip Separation on Tensile Modulus of Boltaron 6200 PVC as Determined from an Instron Testing Machine

Date:	February 19-20, 1963	; May 6, 1963
Temperature:	73°F	; 82°F
Relative Humidity:	Unknown	; Unknown
Specimen:	Test Sample 4-1-A	

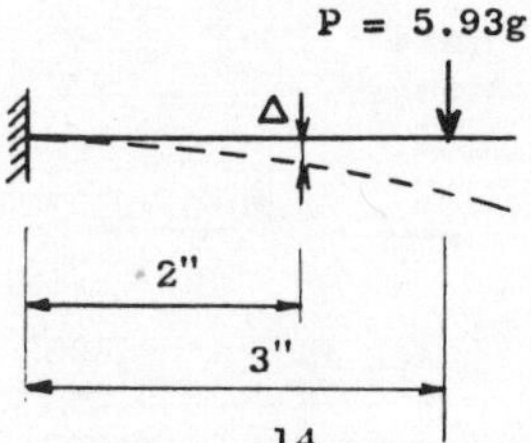

$EI\Delta = \frac{14}{3} P$

Test Sample 4-1-A Width = 0.698"
Thickness = 0.0304"

This coupon was tested a number of times between July 6, 1962 and February 19, 1963 and Δ_{ave} = 0.078"

E = 474,000 psi

Sample was then split longitudinally yielding specimens 0.372" and 0.296" wide.

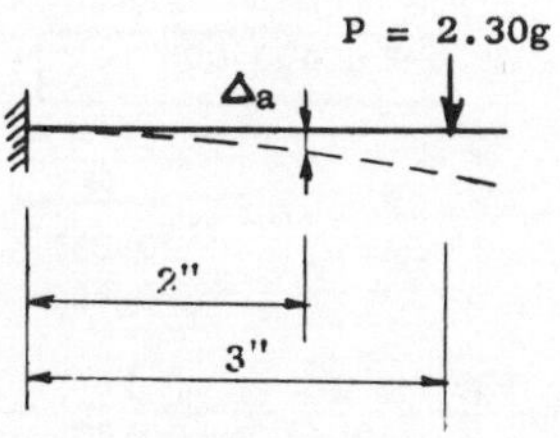

Test Sample 4-1-Aa width = 0.372"
thickness = 0.0306"

		Δ_a	
Run #1	2/20/63	0.057"	
2	"	0.058	
3	"	0.055	
4	"	0.055	
5	5/6/63	0.059	Use 0.057"
6	"	0.058	
7	"	0.056	

E = 467,000 psi

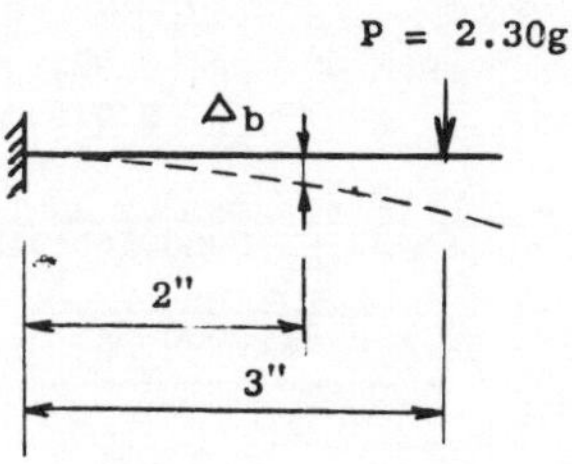

Test Sample 4-1-Ab width = 0.296"
thickness = 0.0302"

		Δ_b	
Run #1	2/20/63	0.080"	
2	"	0.075	
3	"	0.074	
4	"	0.076	Use 0.076"
5	"	0.076	
6	5/6/63	0.076	

E = 457,000 psi

Plate A-6 Effect of Poisson's Ratio on Bending Modulus of Boltaron 6200 PVC as Determined from Cantilever Beam Tests

Date:	March 10, 1963
Temperature:	73°F
Relative Humidity:	Unknown
Specimen:	Cut from as manufactured sheet (i.e. unannealed)

Width = 0.644"
Thickness = 0.0313"

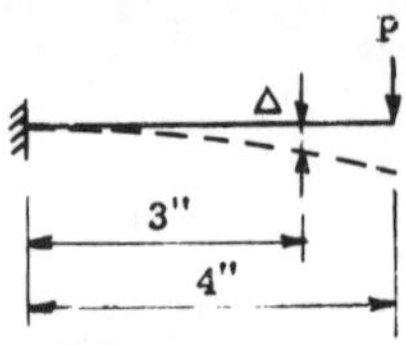

$$EI\Delta = \frac{27}{2}P$$

P = 5.93 Grams

	Δ
Run #1	0.206"
2	0.208
3	0.211
4	0.210

$$E = \frac{27}{2}\ \frac{5.93}{453.6}\ \frac{1}{.00000165}\ \frac{1}{.209}$$

E = 512,000 psi

P = 2.30 Grams

	Δ
Run #5	0.083"
6	0.084
7	0.083

$$E = \frac{27}{2}\ \frac{2.30}{453.6}\ \frac{1}{.00000165}\ \frac{1}{.083}$$

E = 500,000 psi

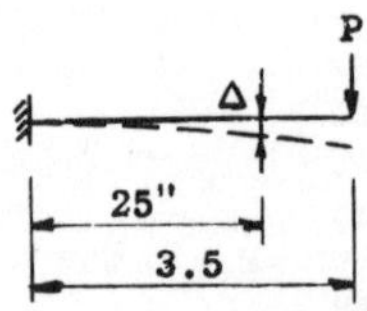

$$EI\Delta = \frac{25}{3}P$$

	Δ
Run #1	0.131"
2	0.132

$$E = \frac{25}{3}\ \frac{5.93}{453.6}\ \frac{1}{.00000165}\ \frac{1}{.143}$$

E = 500,000 psi

	Δ
Run #3	0.053"
4	0.052

$$E = \frac{25}{3}\ \frac{2.30}{453.6}\ \frac{1}{.00000165}\ \frac{1}{.052}$$

E = 492,000 psi

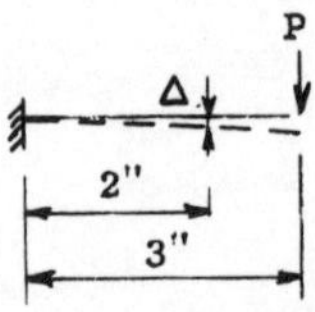

$$EI\Delta = \frac{14}{3}P$$

	Δ
Run #1	0.072"
2	0.073

$$E = \frac{14}{3}\ \frac{5.93}{453.6}\ \frac{1}{.00000165}\ \frac{1}{.073}$$

E = 506,000 psi

	Δ
Run #3	0.027"
4	0.028

$$E = \frac{14}{3}\ \frac{2.30}{453.6}\ \frac{1}{.00000165}\ \frac{1}{.028}$$

E = 511,000 psi

Plate A-7 Effect of Nonlinearity on Bending Modulus of Boltaron 6200 PVC as Determined from Cantilever Beam Tests

Date: May 6, 1963
Temperature: 82°F
Relative Humidity: Unknown
Specimen: Test Sample 4-1-A After Longitudinal Cutting

	4-1-Ab	4-1-Aa
Width	0.296"	0.372"
Thickness	0.0302	0.0306

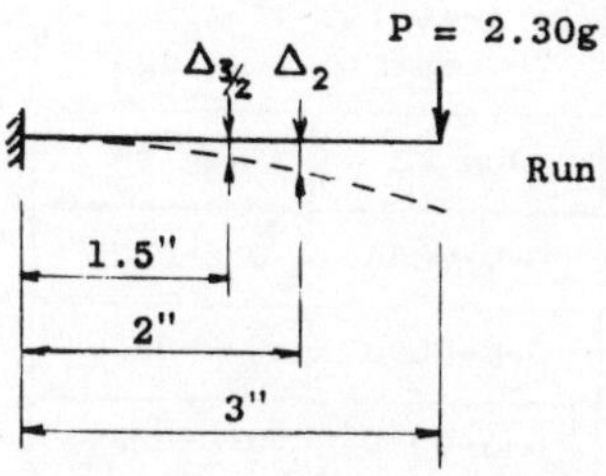

4-1-Ab

	$\Delta_{3/2}$	Δ_2
Run #1	0.046"	0.076"

4-1-Aa

	$\Delta_{3/2}$	Δ_2
Run #1	0.036"	0.059"
2	0.036	0.058
3	0.034	0.056

$$E_{3/2} I \Delta_{3/2} = \frac{45}{16} P$$

$$E_2 I \Delta_2 = \frac{14}{3} P$$

	4-1-Ab	4-1-Aa
$E_{3/2}$	456,000 psi	459,000 psi
E_2	457,000 psi	459,000 psi

These results by themselves are not conclusive. However, with the additional results given in Plate A-6 and the uniformly consistent results obtained throughout the course of the work, it is concluded that an effective fixed support was provided.

Plate A-8 Effect of End Restraint on Bending Modulus of Boltaron 6200 PVC as Determined from Cantilever Beam Tests.

Date: June 2, 1963
Temperature: 80°F
Relative Humidity: Unknown
Specimen: Test Sample 4-1-D   Width = 0.699"
Thickness = 0.0298"

1 vertical and 1 horizontal gage attached to each side

Gage: Baldwin-Lima-Hamilton   FA - 25 - 12 S-6
G.F. = 2.03 ± 1%
R = 120.0 ± 0.2 Ω

Cement: Baldwin-Lima-Hamilton EP 150

Load	Stress	Vertical Gages In Arms 1&3 of a full Wheatstone Bridge		Horizontal Gages In Arms 1&3 of a full Wheatstone Bridge	
		Run #1	Run #2	Run #3	Run #4
zero	0	0-6-1600	0-6-1645	0-8-0850	0-8-0825
5 lbs.	240 psi	0-8-0280	0-8-0345	0-8-0590	0-8-0550
10	480	0-8-0980	0-8-1045	0-8-0310	0-8-0280
15	721	0-8-1700	0-8-1765	0-8-0035	0-8-0020
10		0-8-1000	0-8-1065	0-8-0295	0-8-0290
5		0-8-0300	0-8-0360	0-8-0560	0-8-0550
zero		0-6-1610	0-6-1660	0-8-0830	0-8-0825

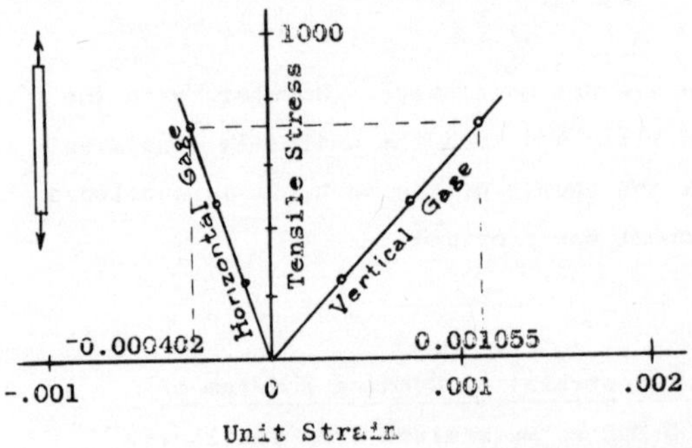

E = 683,000 psi
ν = 0.38

Plate A-9 SR-4 Gage Tests for Poisson's Ratio of Boltaron 6200 PVC

Specimens:	Test Samples	5-3-A	and	5-3-B
	Thickness	0.0290"		0.0298"
	Width	0.695"		0.696"

Standard Tests

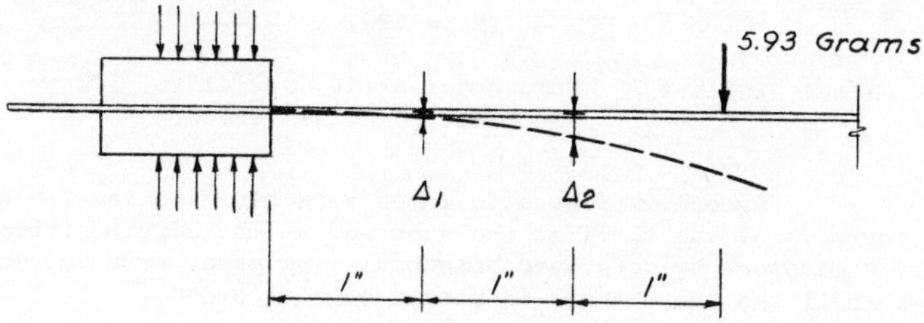

	5-3-A Δ_1	5-3-A Δ_2	5-3-B Δ_1	5-3-B Δ_2
7/31/62 @ 76°F		0.089"		0.079"
5/15/63 @ 76°F	0.027"	0.095"	0.022"	0.080"

Tests to Simulate Shell Edge Fixity

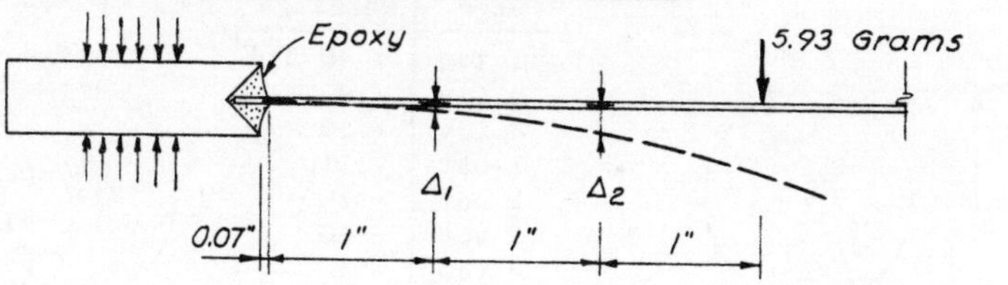

	5-3-A Δ_1	5-3-A Δ_2	5-3-B Δ_1	5-3-B Δ_2
5/15/63 @ 76°F	0.025"	0.093"	0.023"	0.083"
Removed and Reglued				
5/31/63 @ 81°F	0.026"	0.088"	0.022"	0.077"

No Increase in 2 Minutes

Plate A-10 Test Data for Determining Degree of Model Shell Edge Restraint

Date:	July 24-25, 1963
Temperature:	85-90°F
Relative Humidity:	Unknown
Specimen:	Shell 6-1
Gages:	BLH AR-2 Rosettes

$$G.F. = 2.05 \pm 1\%$$

$$R = 120.0 \pm 0.5\,\Omega$$

Strain Indicator: Budd Model A-110 Serial No. 172
10 Channel, digital readout

Procedure: Rosette gages were glued to the top and bottom surfaces of the shell at the apex and at one edge position. The half-Wheatstone bridges were balanced. The gages were cut out from the shell and the changes in strain were recorded.

With only one dummy gage, there was a gage heating problem, and it was necessary to make the initial balances and subsequent readings after two-minute warming periods. The drift in this time was about 30 micro-inches/inch.

Initial balances were made with the shell edge resting on a table top. The actual readings were taken with the shell segment suspended by the lead wires. Different means of suspension changed readings by up to 30 micro-inches/inch. The recorded values are averages.

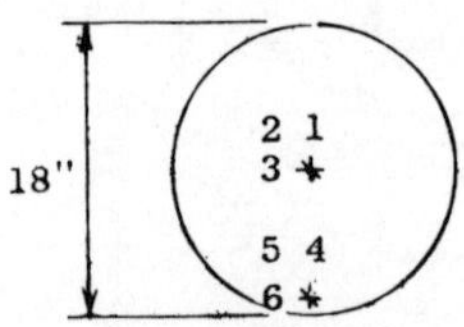

Gage Plan

Gage	Strain Readings
1 top	+ 40 u"/"
2 top	+ 90
3 top	+ 30
1 bot	- 110
2 bot	- 100
3 bot	- 10
4 top	+ 10
5 top	- 15
6 top	- 30
4 bot	- 60
5 bot	- 15
6 bot	+ 20

E = 450,000 psi

$\nu = 0.38$

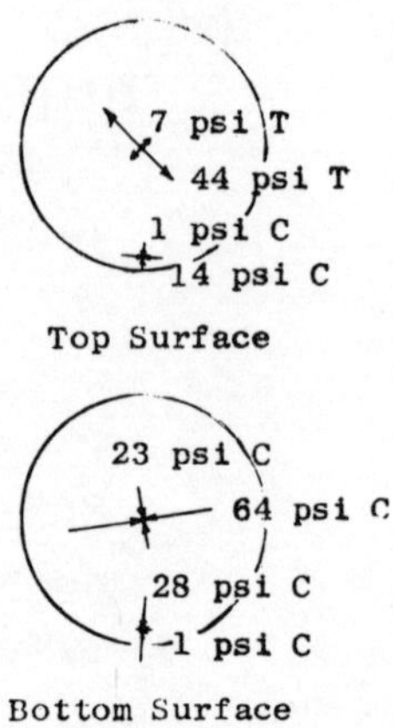

PRINCIPAL STRESSES

Plate A-11 Residual Stresses in Model Shell 6-1

APPENDIX B

Data for Air-Pressure-Loaded 18-Inch Radius Shells

Series 1. 18-Inch Radius Spherical Domes

Mold: Male wooden mold manufactured by F. W. Dixon Co., Cambridge, Massachusetts, at a cost of $75. Hardwood mold was ordered, but through their error it was made of white pine. To kill the grain they resanded the mold and finished the surface with an epoxy paint. The mold was extremely smooth but contained a relatively flat spot at the top.

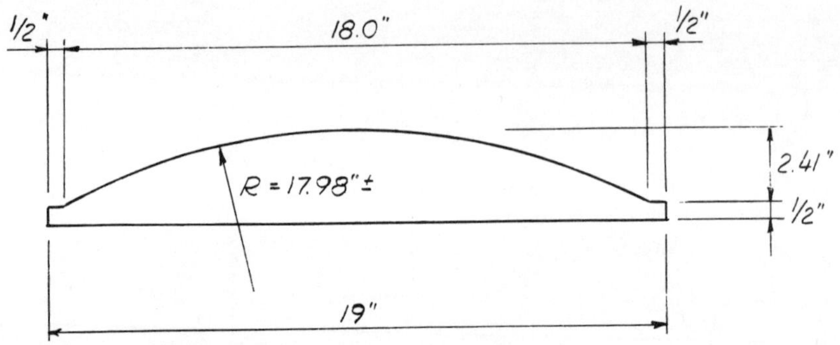

Plastic: Boltaron 6200 PVC purchased April 2, 1962.

Vacuum-Forming: Three models formed June 13, 1962, on the machine in the Laboratory for Structural Models.

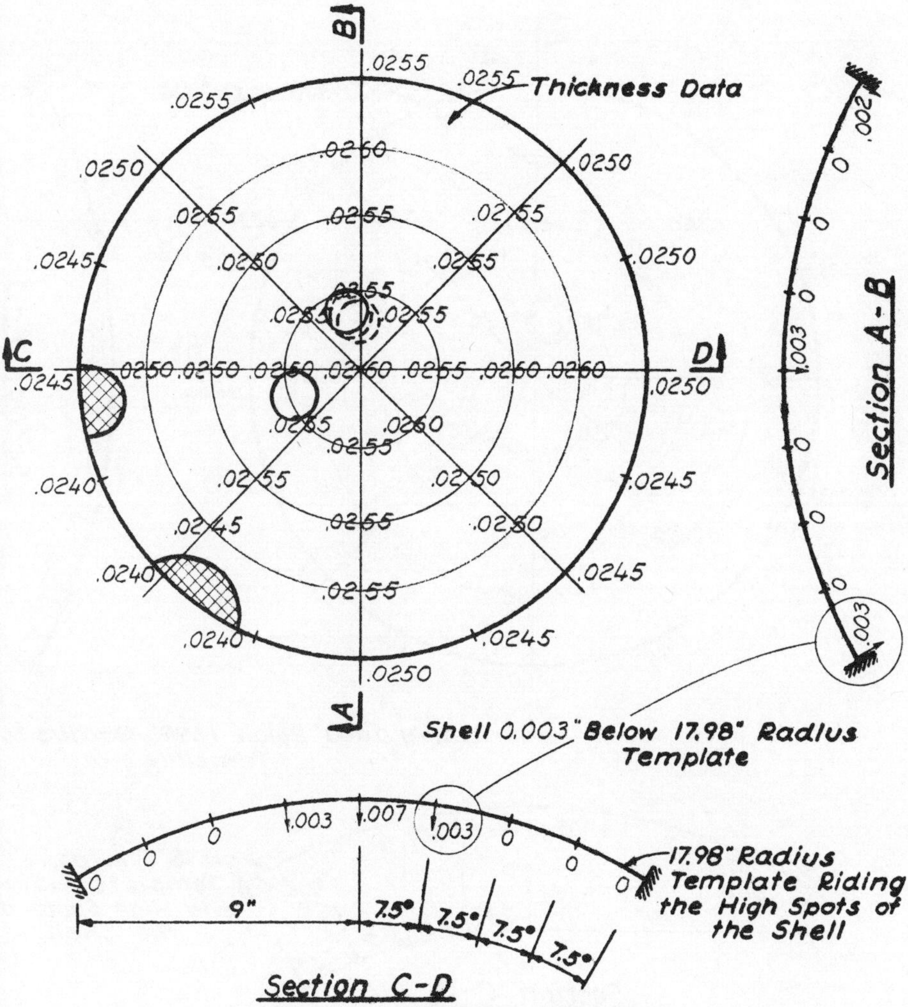

Maximum Thickness = 0.0260"
Minimum Thickness = 0.0240
Average Thickness = 0.0253
Thickness @ Air Pressure Buckle Position = 0.0255

Air Pressure Buckle Position

Air Pressure Buckle Positions of Shells 1-2 and 1-3

Hanging Weights Buckle Position

Data Sheet for Shell 1-1

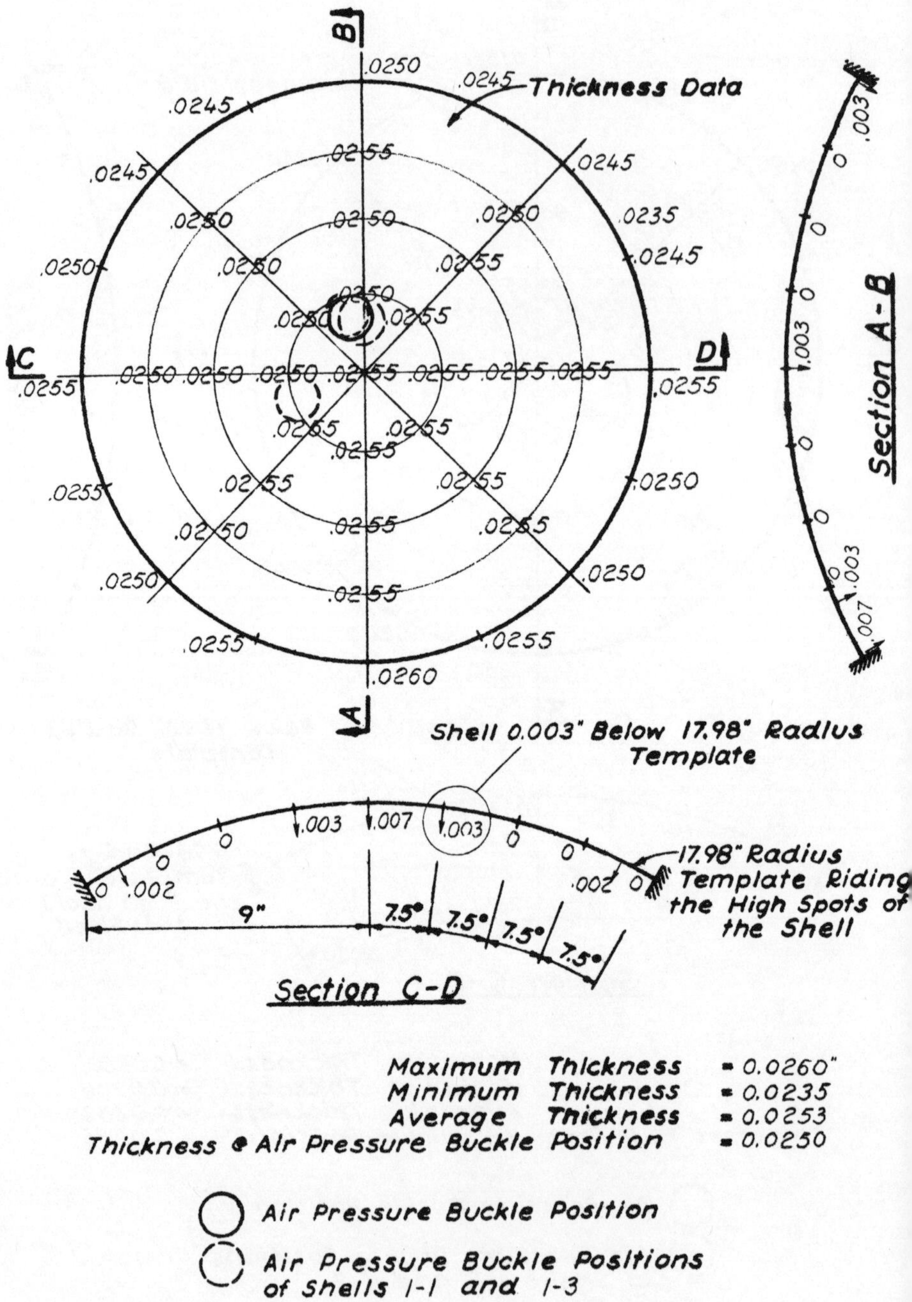

Data Sheet for Shell 1-2

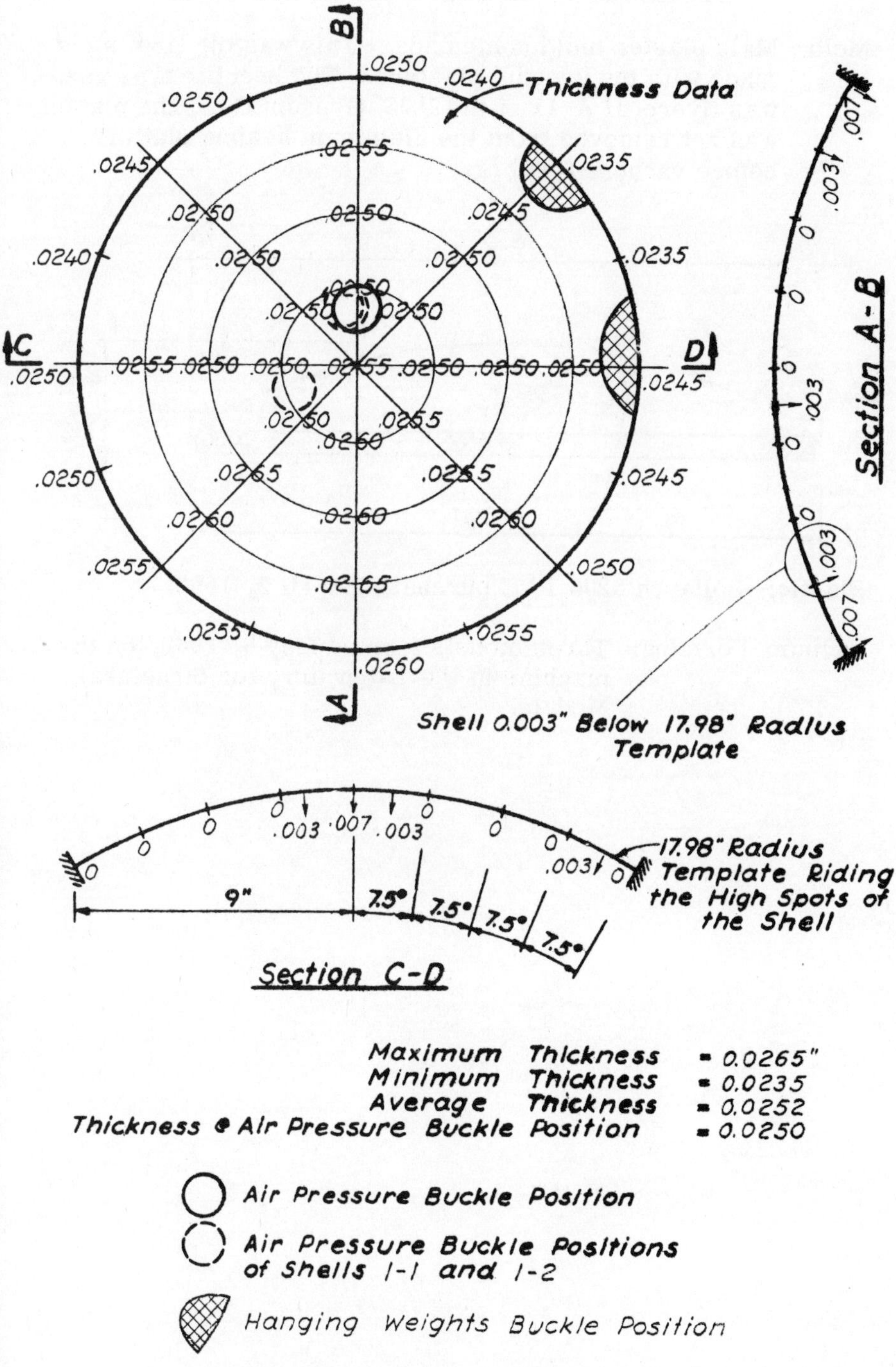

Data Sheet for Shell 1-3

Series 2. 18-Inch Radius Spherical Domes

Mold: Male plaster mold homemade. This was the first mold made with the gypsum plasters. The specific type used was Hydrocal A-11 of the U. S. Gypsum Co. The plaster was not removed from the aluminum holding platform before vacuum-forming.

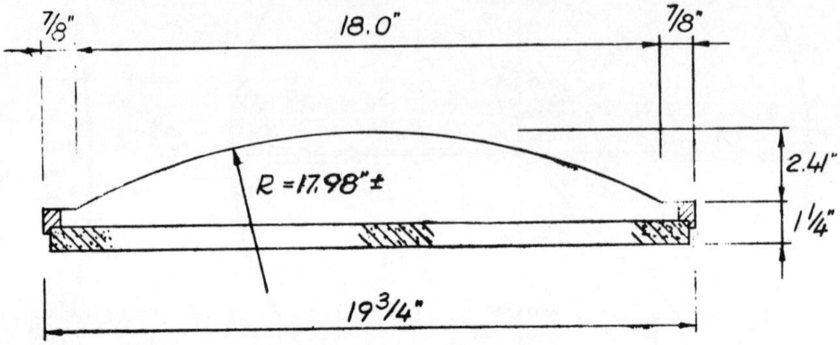

Plastic: Boltaron 6200 PVC purchased April 2, 1962.

Vacuum-Forming: Three models formed May 9, 1962, on the machine in the Laboratory for Structural Models.

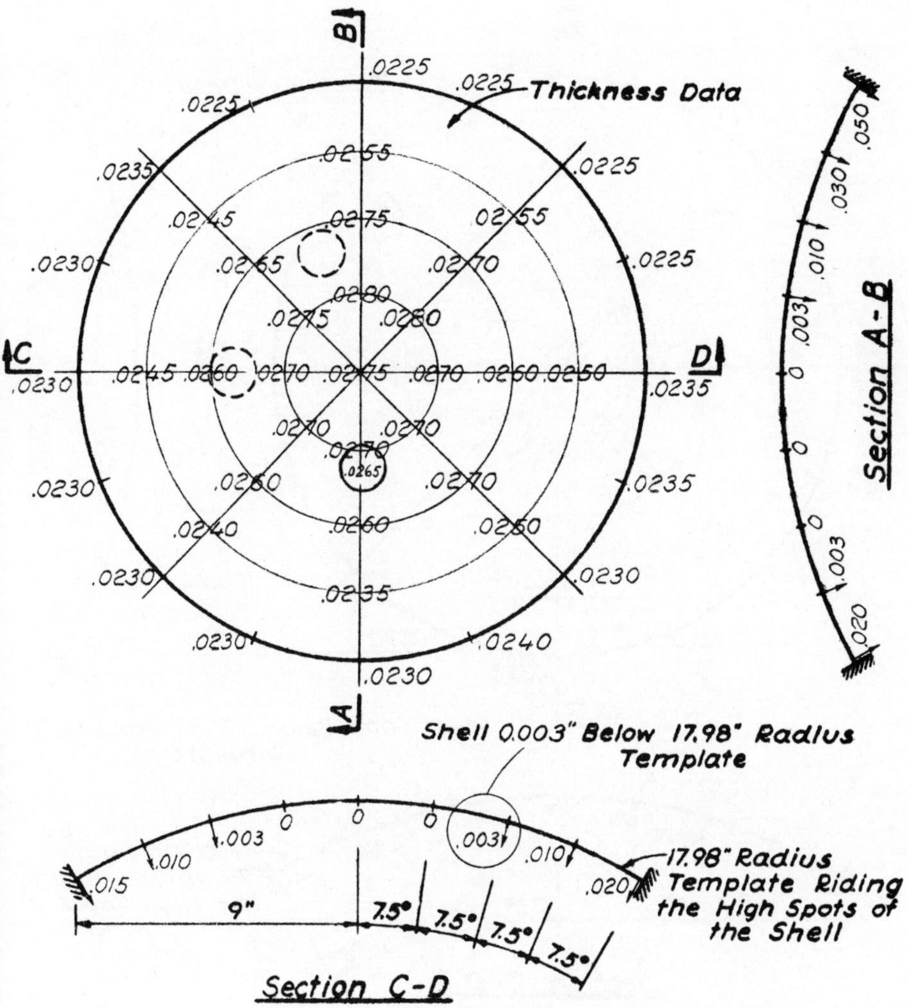

Maximum	Thickness	= 0.0275"
Minimum	Thickness	= 0.0225
Average	Thickness	= 0.0252
Thickness @ Air Pressure	Buckle Position	= 0.0265

○ Air Pressure Buckle Position

◌ Air Pressure Buckle Positions of Shells 2-2 and 2-3

Data Sheet for Shell 2-1

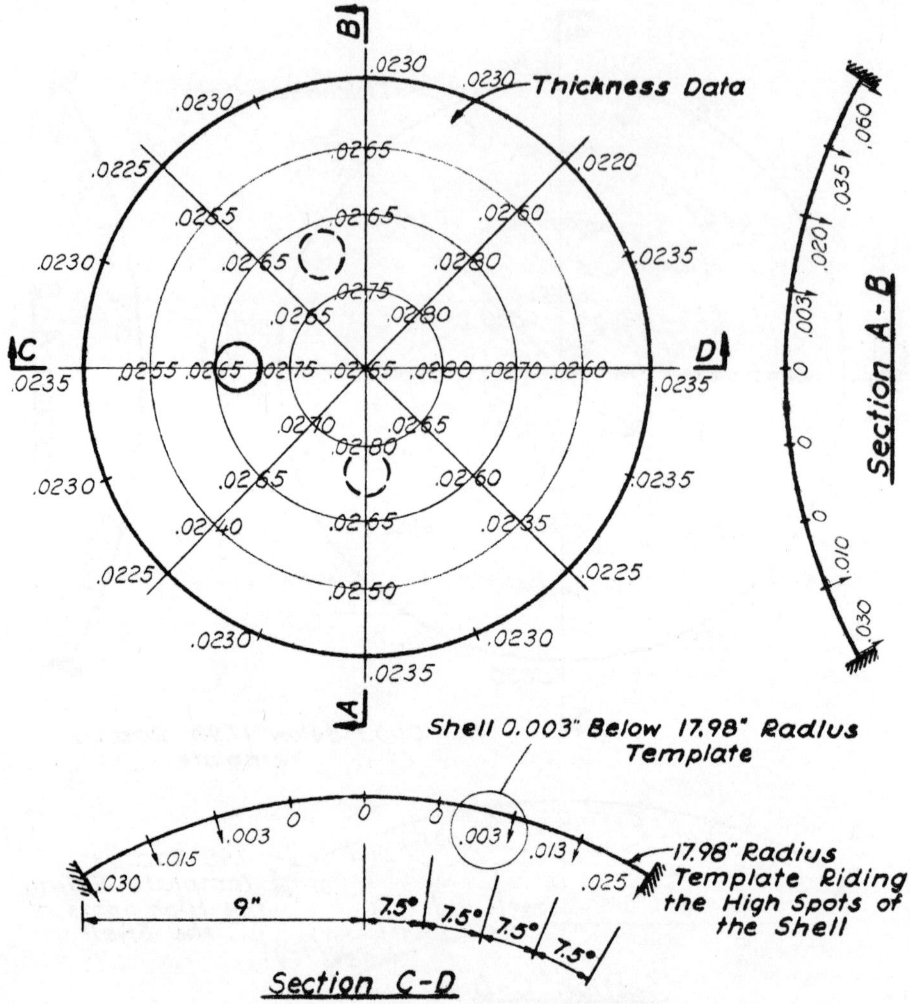

Maximum Thickness	= 0.0280"
Minimum Thickness	= 0.0220
Average Thickness	= 0.0254
Thickness @ Air Pressure Buckle Position	= 0.0270

◯ Air Pressure Buckle Position

◌ Air Pressure Buckle Positions of Shells 2-1 and 2-3

Data Sheet for Shell 2-2

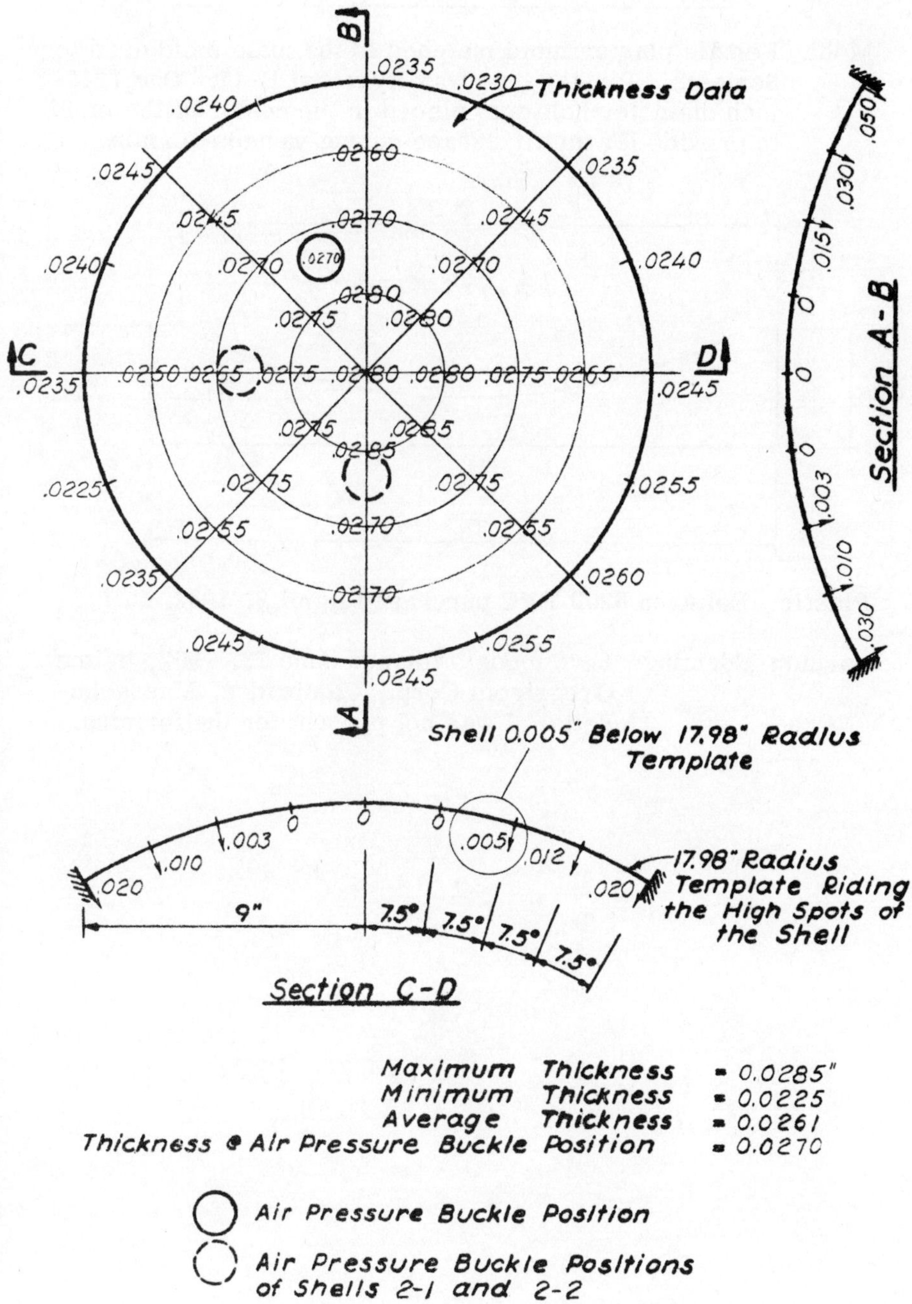

Data Sheet for Shell 2-3

Series 3. 18-Inch Radius Spherical Domes

Mold: Female plaster mold matched to the male mold used for Series 2. Plaster type was Hydrocal B-11. One 1/16-inch diameter hole was placed in the center of the mold to provide for an air escape during vacuum-forming.

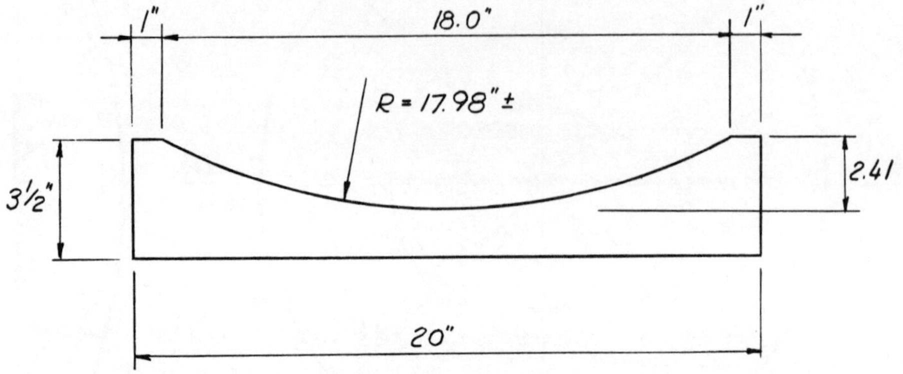

Plastic: Boltaron 6200 PVC purchased April 2, 1962.

Vacuum-Forming: Five models formed June 13, 1962, by the Gregstrom Corp., Cambridge, Massachusetts. I was not present for the forming.

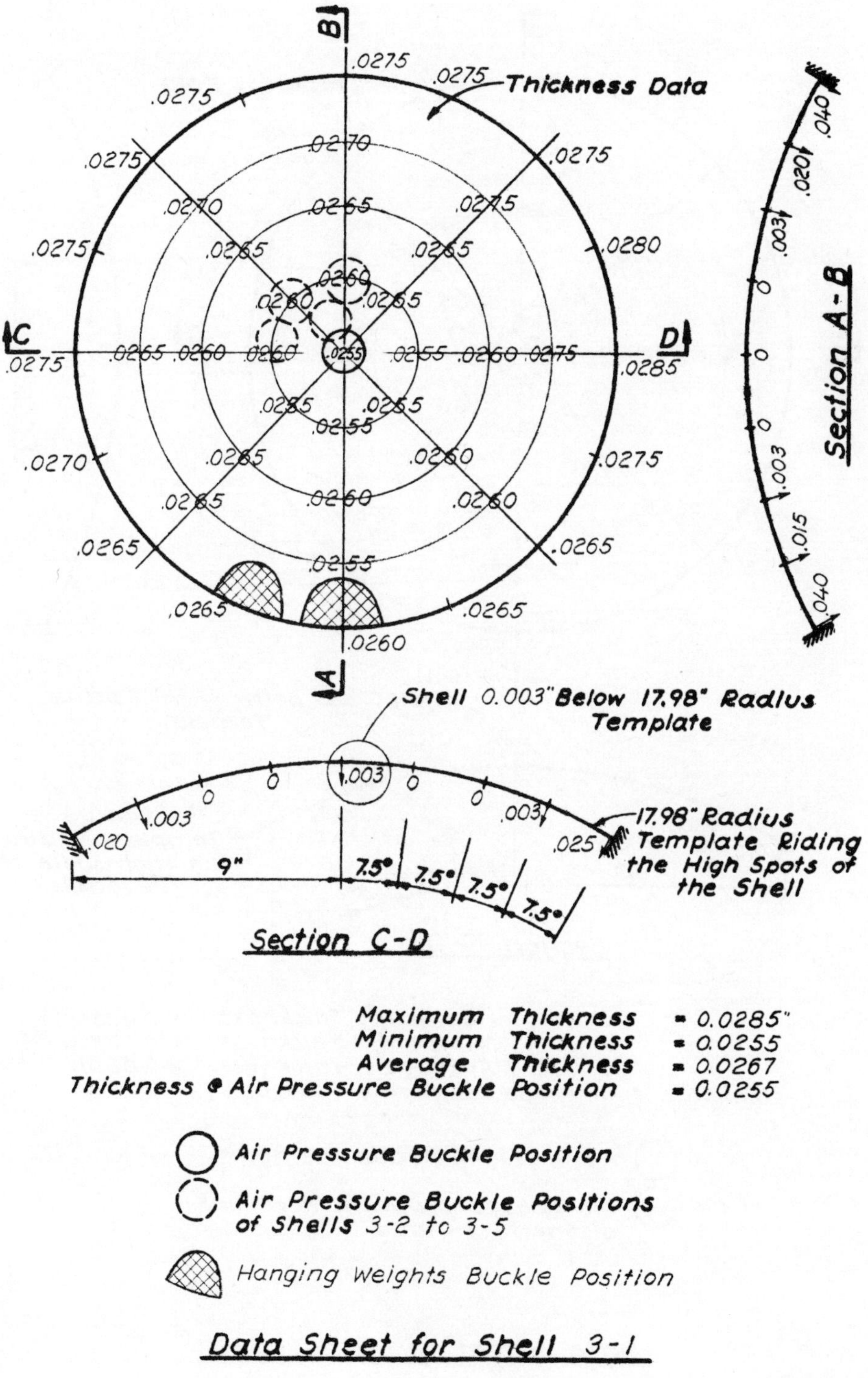

Data Sheet for Shell 3-1

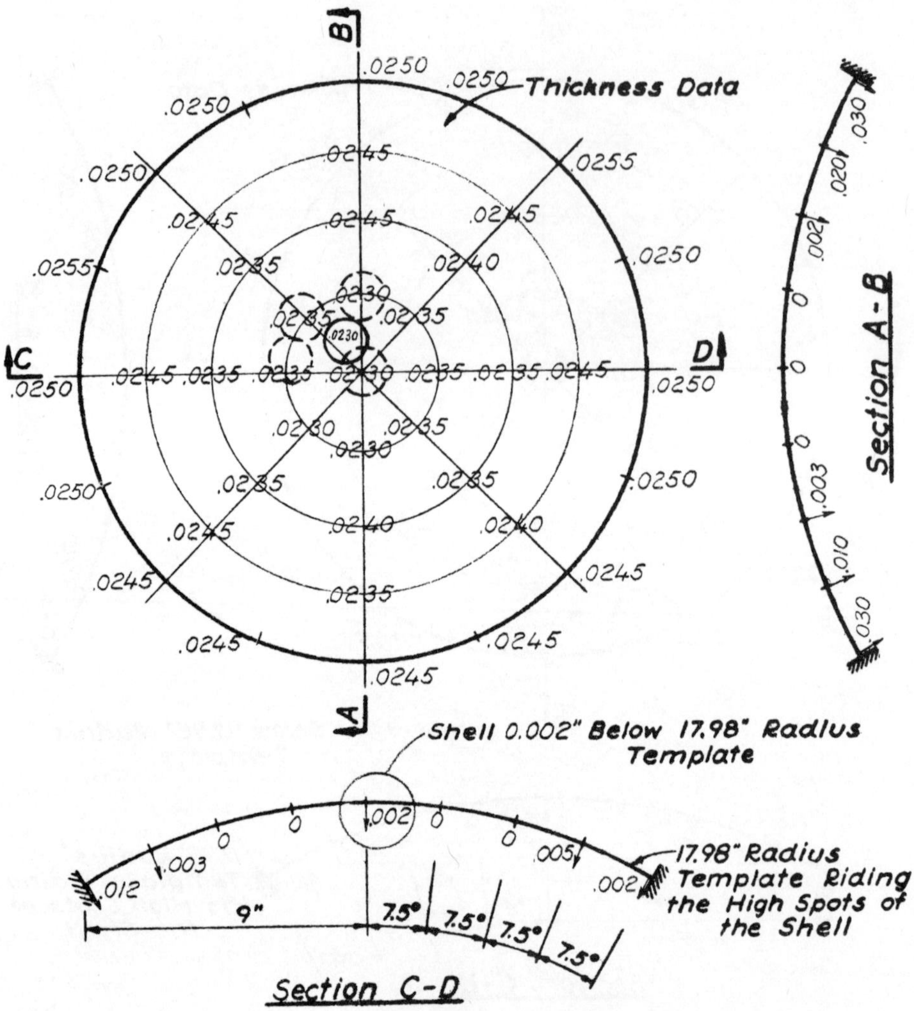

Maximum Thickness = 0.0255"
Minimum Thickness = 0.0230
Average Thickness = 0.0241
Thickness @ Air Pressure Buckle Position = 0.0230

Air Pressure Buckle Position

Air Pressure Buckle Positions of Shells 3-1 and 3-3 to 3-5

Data Sheet for Shell 3-2

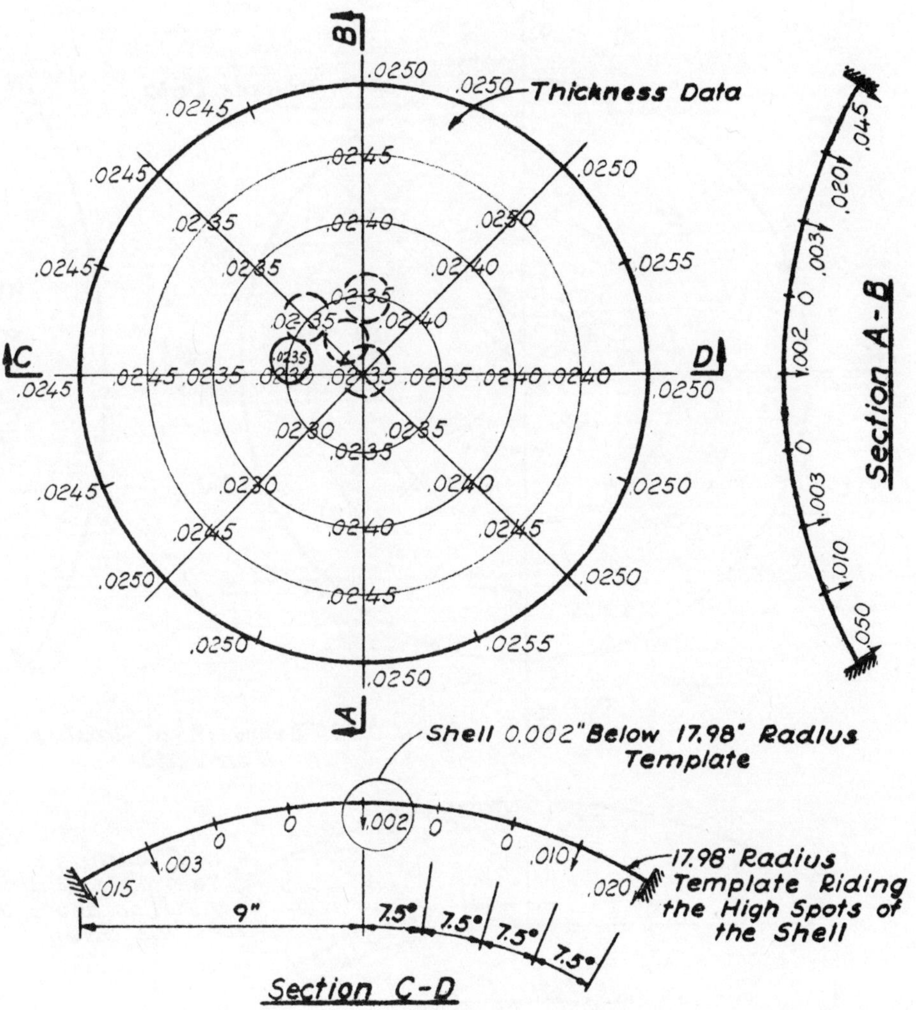

Maximum	Thickness	= 0.0255"
Minimum	Thickness	= 0.0230
Average	Thickness	= 0.0242
Thickness @ Air Pressure Buckle Position		= 0.0235

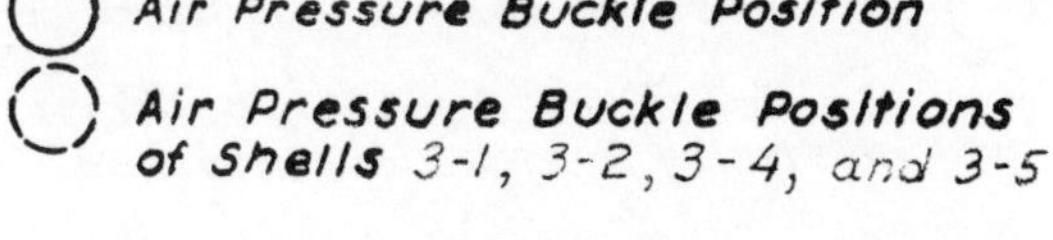

Data Sheet for Shell 3-3

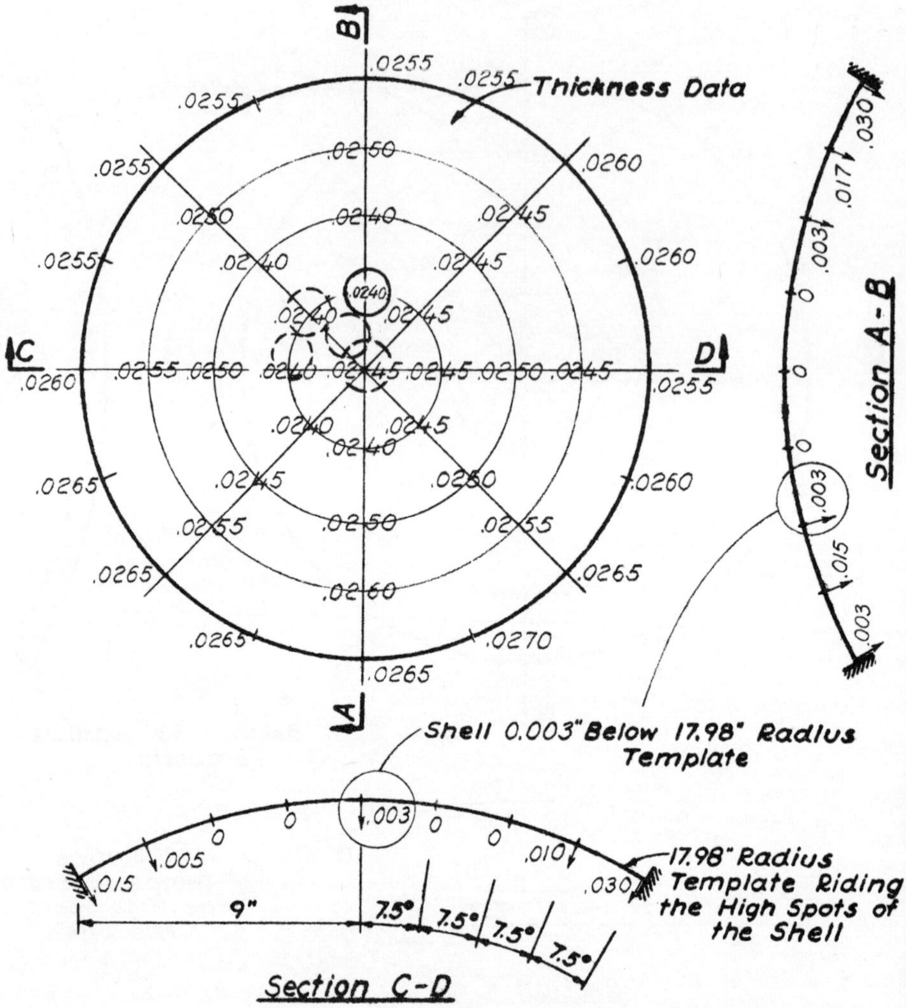

Maximum Thickness = 0.0270"
Minimum Thickness = 0.0240
Average Thickness = 0.0251
Thickness @ Air Pressure Buckle Position = 0.0240

Solid circle: Air Pressure Buckle Position

Dashed circle: Air Pressure Buckle Positions of Shells 3-1, 3-2, 3-3 and 3-5

Data Sheet for Shell 3-4

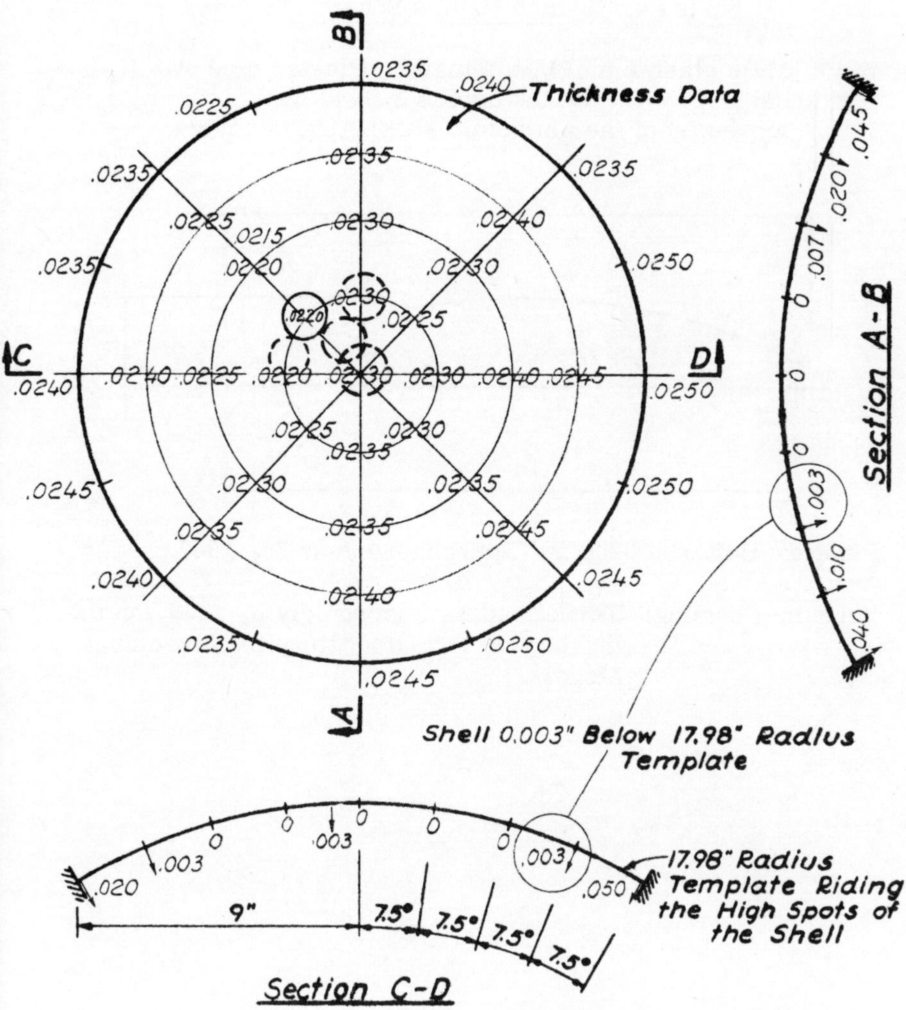

Maximum Thickness = 0.0250"
Minimum Thickness = 0.0215
Average Thickness = 0.0239
Thickness @ Air Pressure Buckle Position = 0.0220

Air Pressure Buckle Position

Air Pressure Buckle Positions of Shells 3-1 to 3-4

Data Sheet for Shell 3-5

Series 4. 18-Inch Radius Spherical Domes

Mold: Male plaster mold homemade. Plaster type was Hydrocal B-11. Unlike the Series 2 mold the plaster was removed from the aluminum supporting elements.

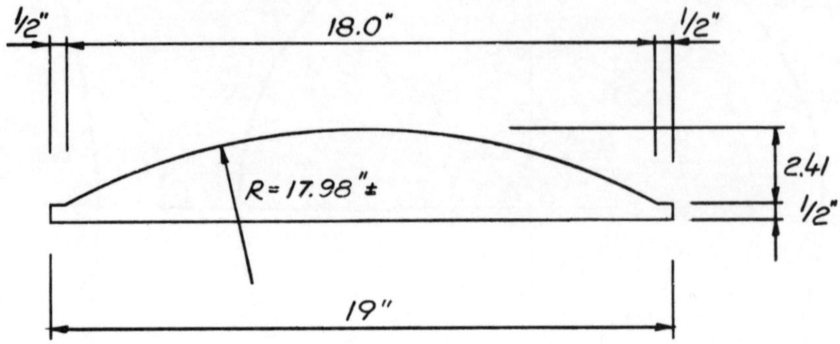

Plastic: Boltaron 6200 PVC purchased June 29, 1962.

Vacuum-Forming: Three models formed July 3, 1962, on the machine in the Laboratory for Structural Models.

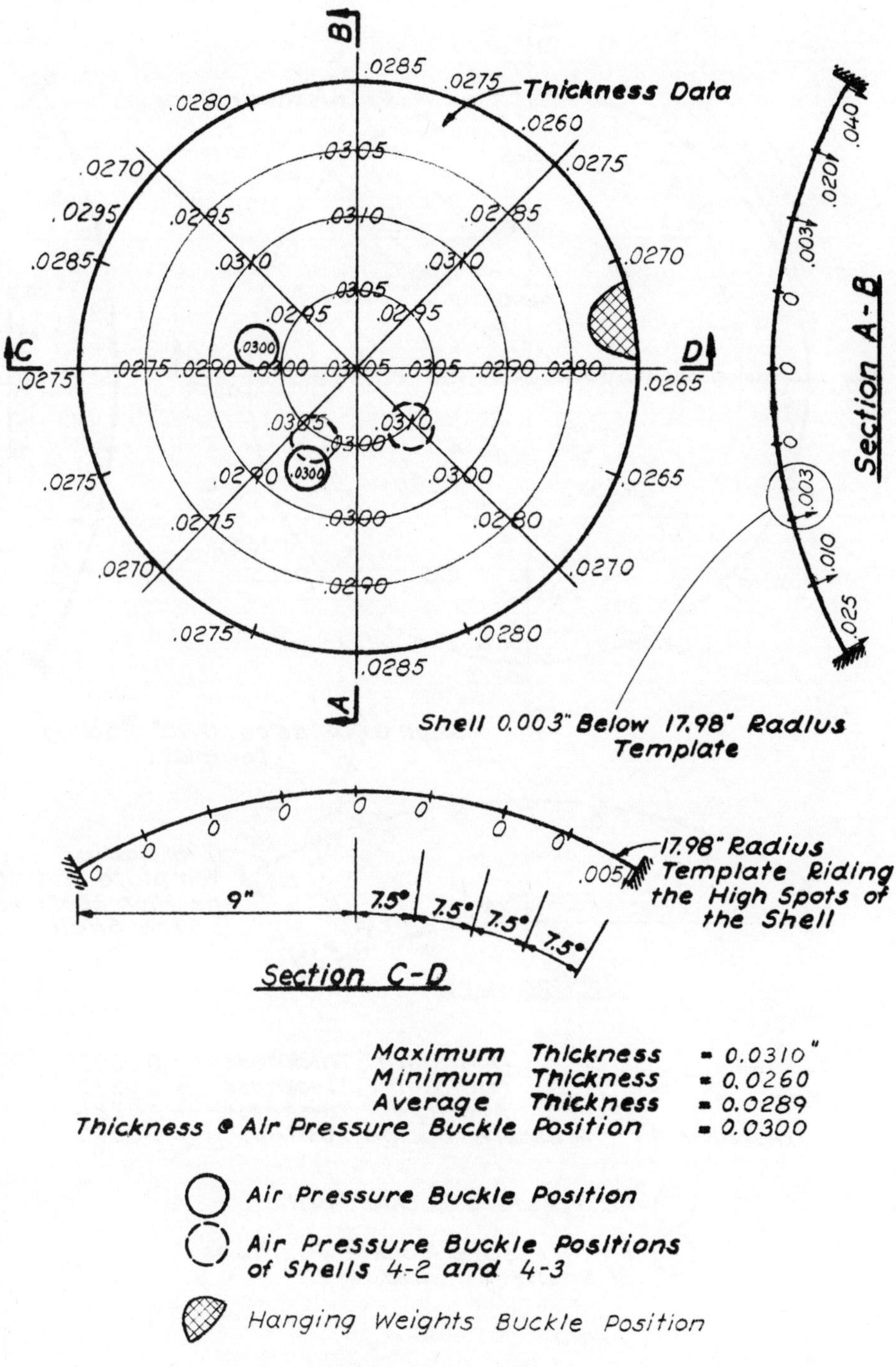

Data Sheet for Shell 4-1

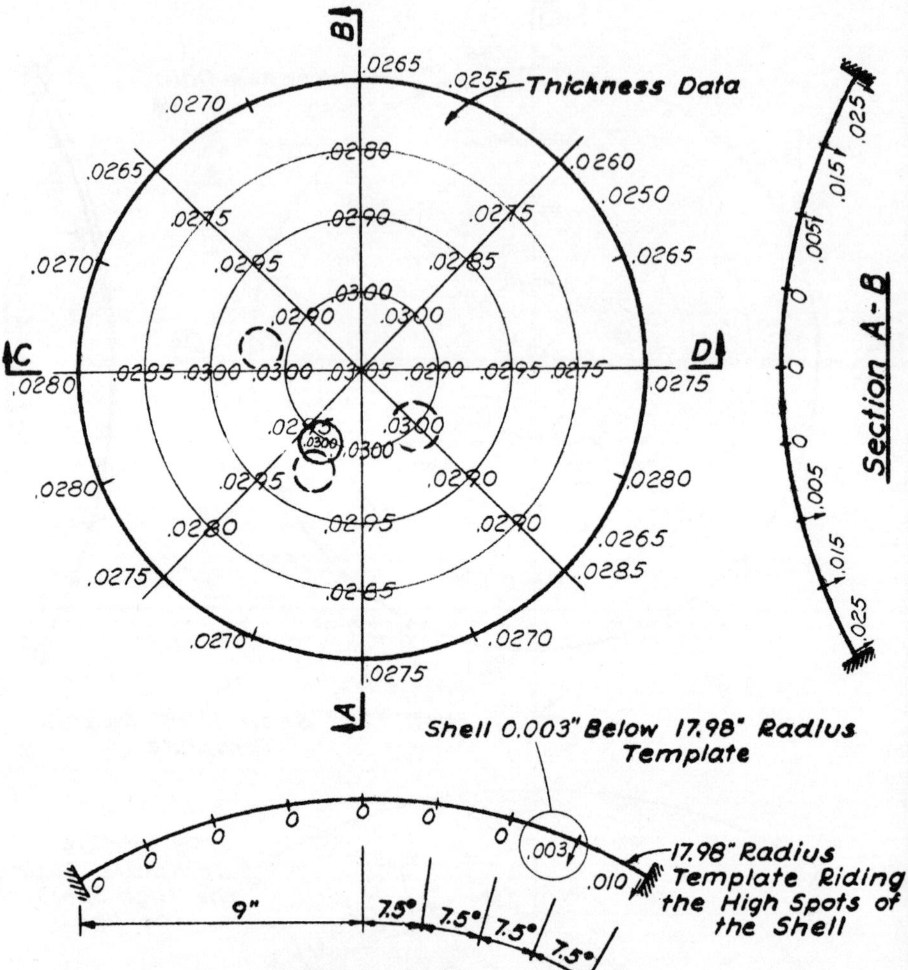

Maximum	Thickness	= 0.0305"
Minimum	Thickness	= 0.0250
Average	Thickness	= 0.0284
Thickness @ Air Pressure Buckle	Position	= 0.0300

◯ Air Pressure Buckle Position

◌ Air Pressure Buckle Positions of Shells 4-1 and 4-3

Data Sheet for Shell 4-2

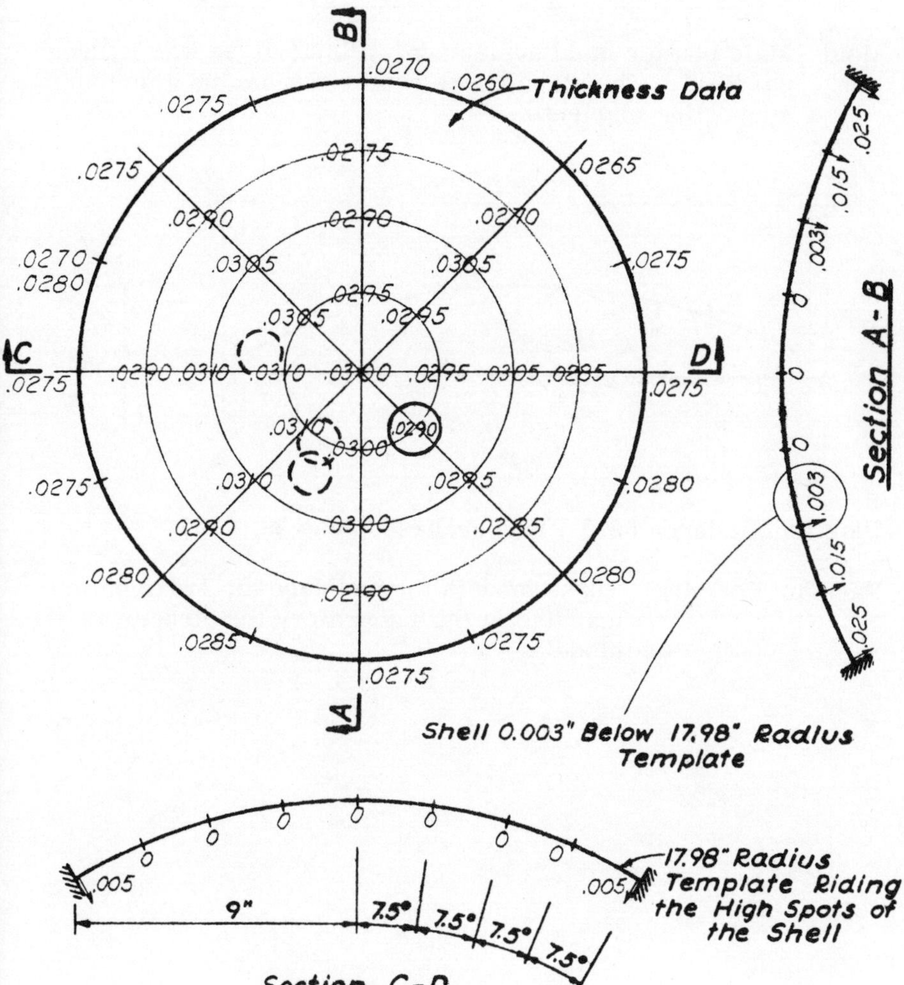

Maximum Thickness	=	0.0310"
Minimum Thickness	=	0.0260
Average Thickness	=	0.0289
Thickness @ Air Pressure Buckle Position	=	0.0290

Air Pressure Buckle Position

Air Pressure Buckle Positions of Shells 4-1 and 4-2

Data Sheet for Shell 4-3

Series 5. 18-Inch Radius Spherical Domes

Mold: Male plaster mold homemade. Plaster type was Hydrocal B-11. The plaster was removed from the aluminum supporting elements.

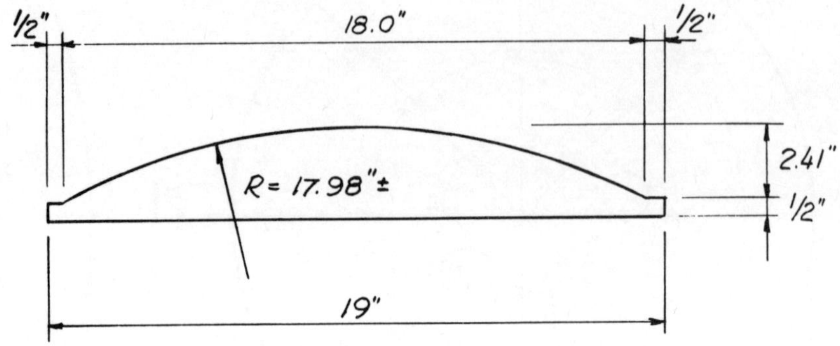

Plastic: Boltaron 6200 PVC purchased June 29, 1962.

Vacuum-Forming: Three models formed July 25, 1962, on the machine in the Laboratory for Structural Models.

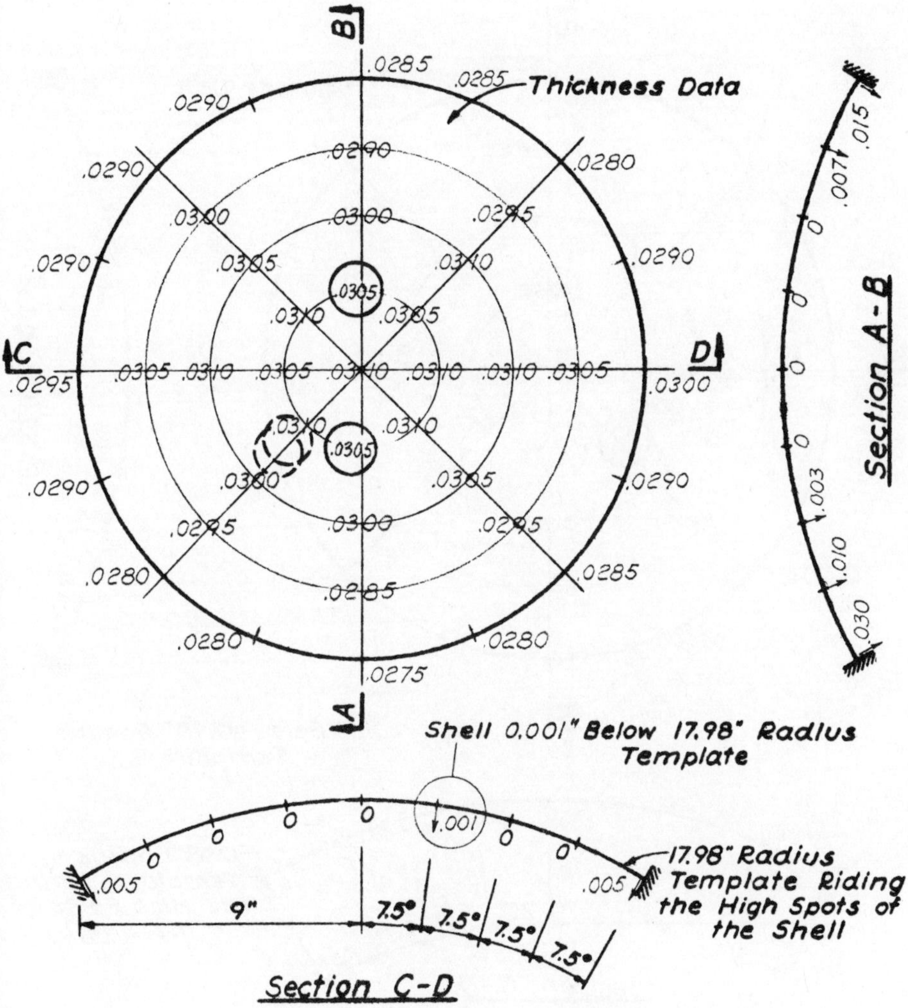

Maximum Thickness = 0.0310"
Minimum Thickness = 0.0275
Average Thickness = 0.0298
Thickness @ Air Pressure Buckle Position = 0.0305

Air Pressure Buckle Position

Air Pressure Buckle Positions of Shells 5-2 and 5-3

Data Sheet for Shell 5-1

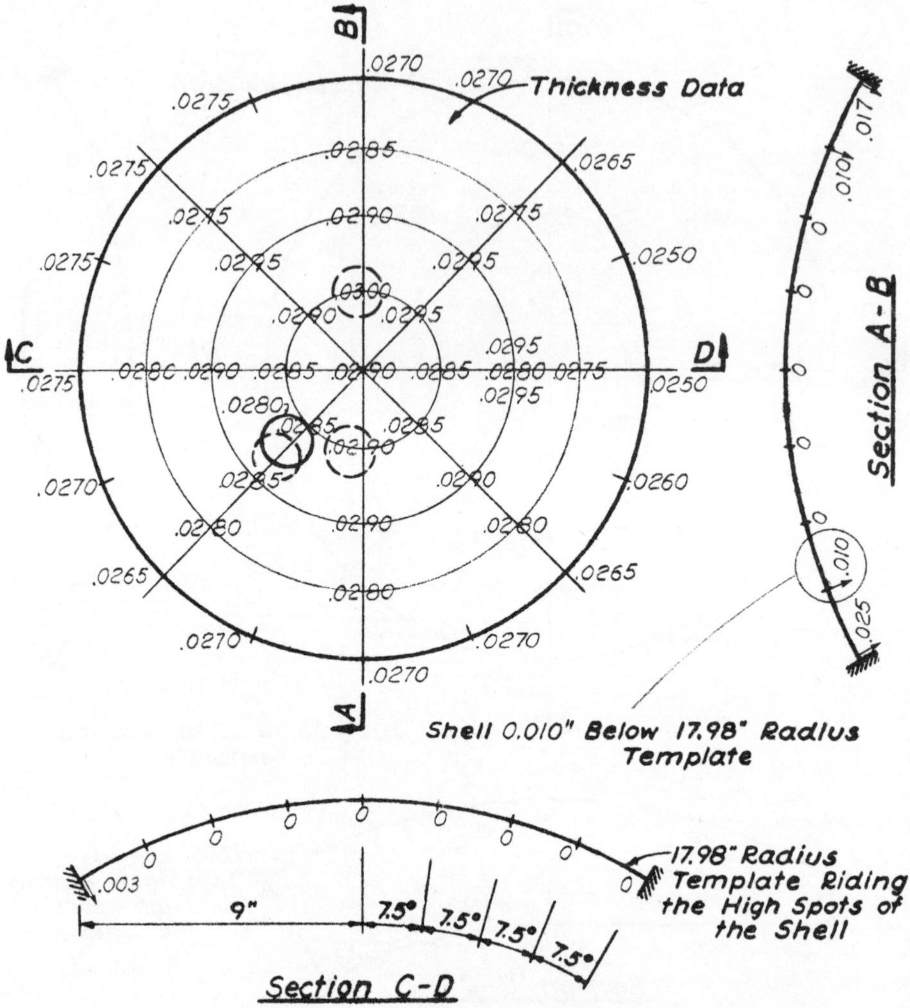

Maximum Thickness = 0.0300"
Minimum Thickness = 0.0250
Average Thickness = 0.0281
Thickness @ Air Pressure Buckle Position = 0.0280

Air Pressure Buckle Position

Air Pressure Buckle Positions of Shells 5-1 and 5-3

Data Sheet for Shell 5-2

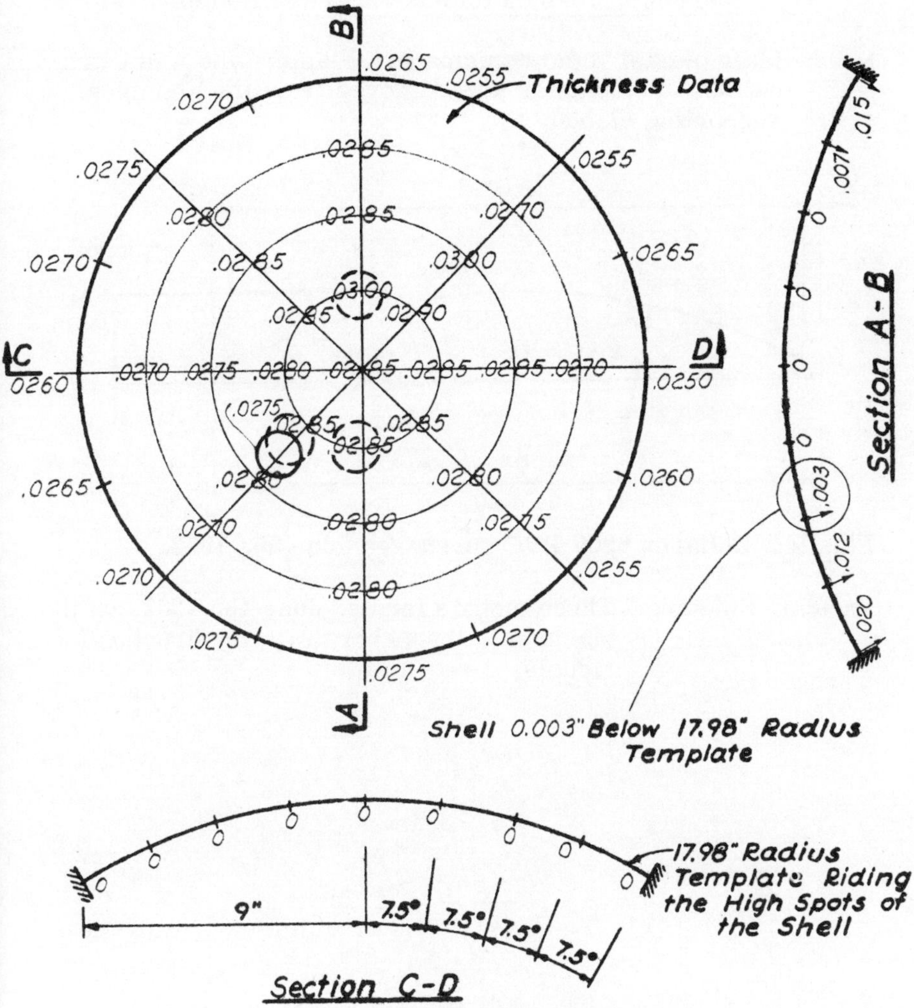

Maximum Thickness	= 0.0300"
Minimum Thickness	= 0.0250
Average Thickness	= 0.0277
Thickness @ Air Pressure Buckle Position	= 0.0275

○ Air Pressure Buckle Position

◌ Air Pressure Buckle Positions of Shells 5-1 and 5-2

Data Sheet for Shell 5-3

Series 6. 18-Inch Radius Spherical Domes

Mold: Male plaster mold homemade. Plaster type was Ultracal 30. The plaster was removed from the aluminum supporting elements.

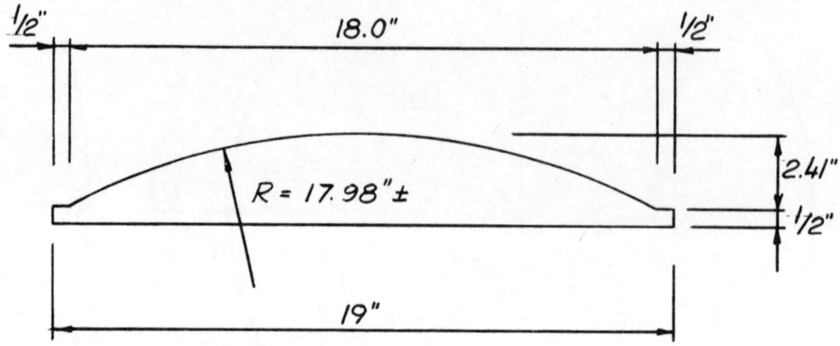

Plastic: Boltaron 6200 PVC purchased June 29, 1962.

Vacuum-Forming: Three models formed June 18, 1963, on the machine in the Laboratory for Structural Models.

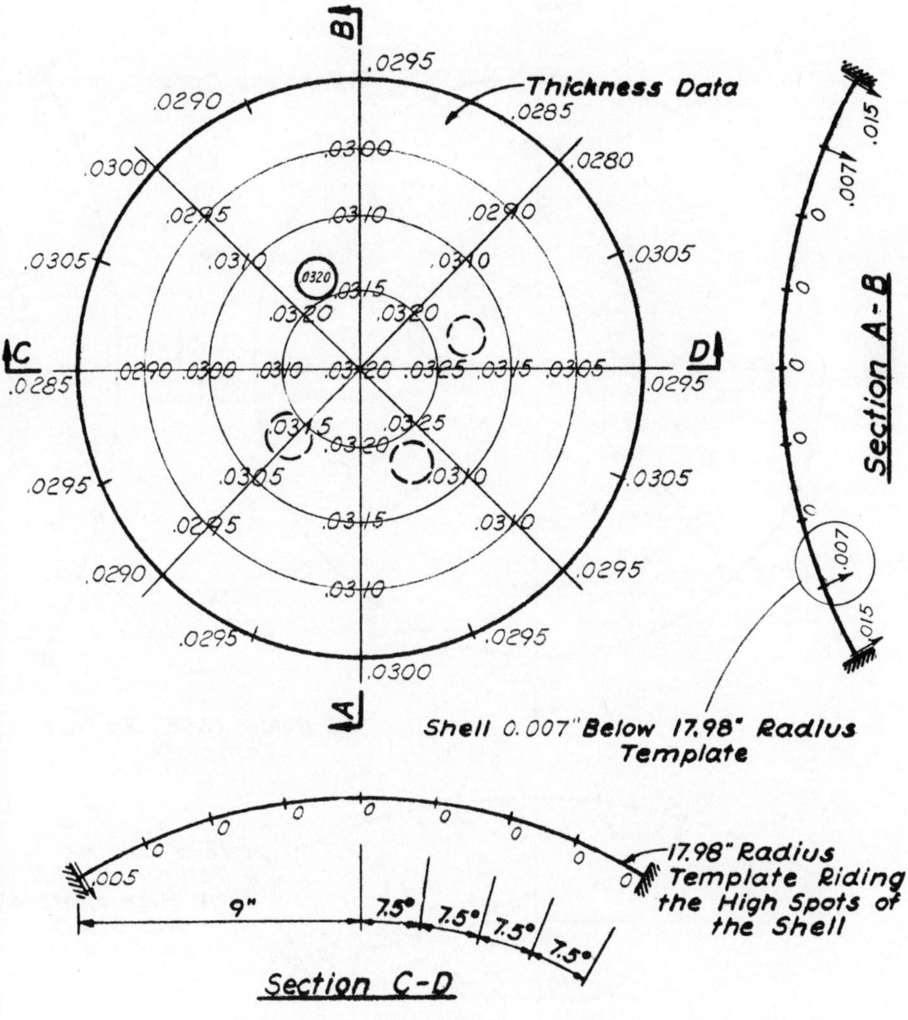

Maximum	Thickness	= 0.0325"
Minimum	Thickness	= 0.0280
Average	Thickness	= 0.0304
Thickness @ Air Pressure Buckle Position		= 0.0320

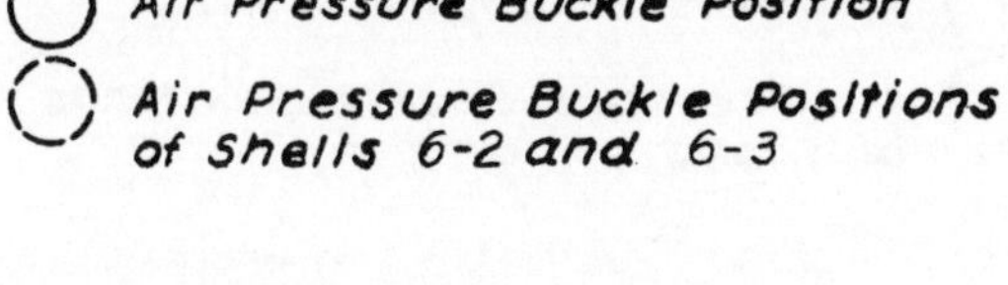

Data Sheet for Shell 6-1

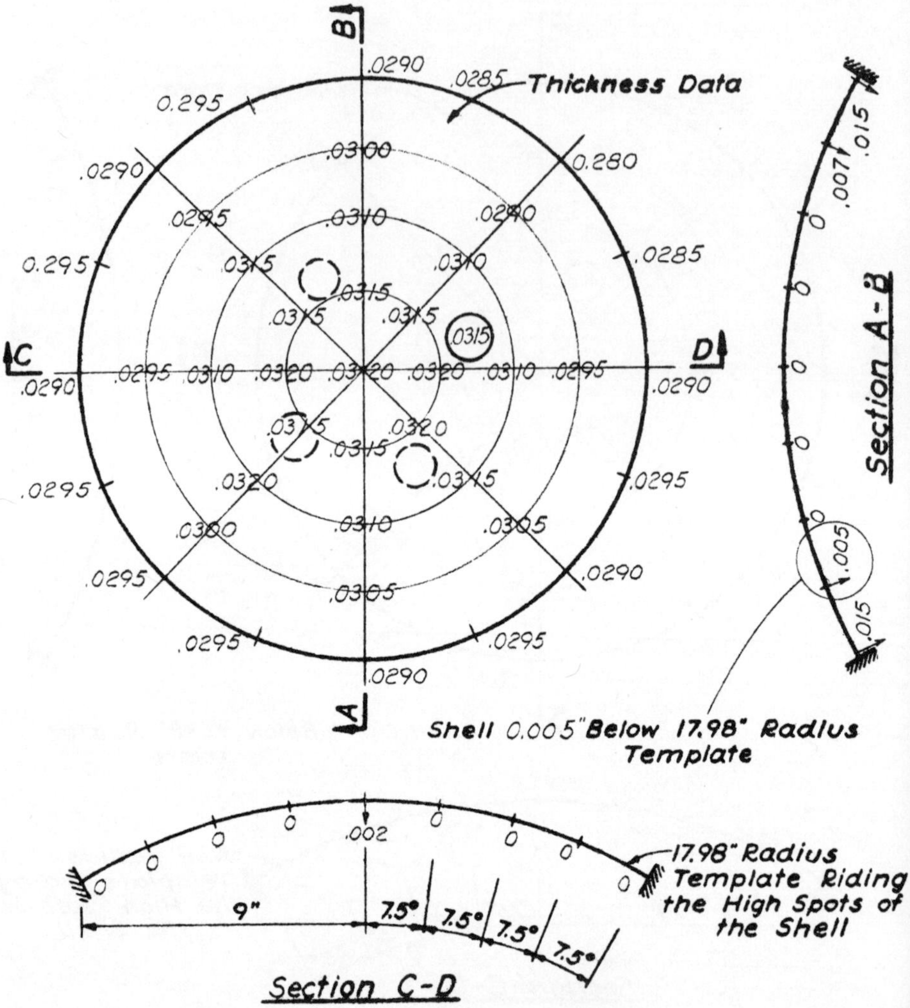

Maximum Thickness = 0.0320"
Minimum Thickness = 0.0280
Average Thickness = 0.0303
Thickness @ Air Pressure Buckle Position = 0.0315

○ Air Pressure Buckle Position

◌ Air Pressure Buckle Positions of Shells 6-1 and 6-3

Data Sheet for Shell 6-2

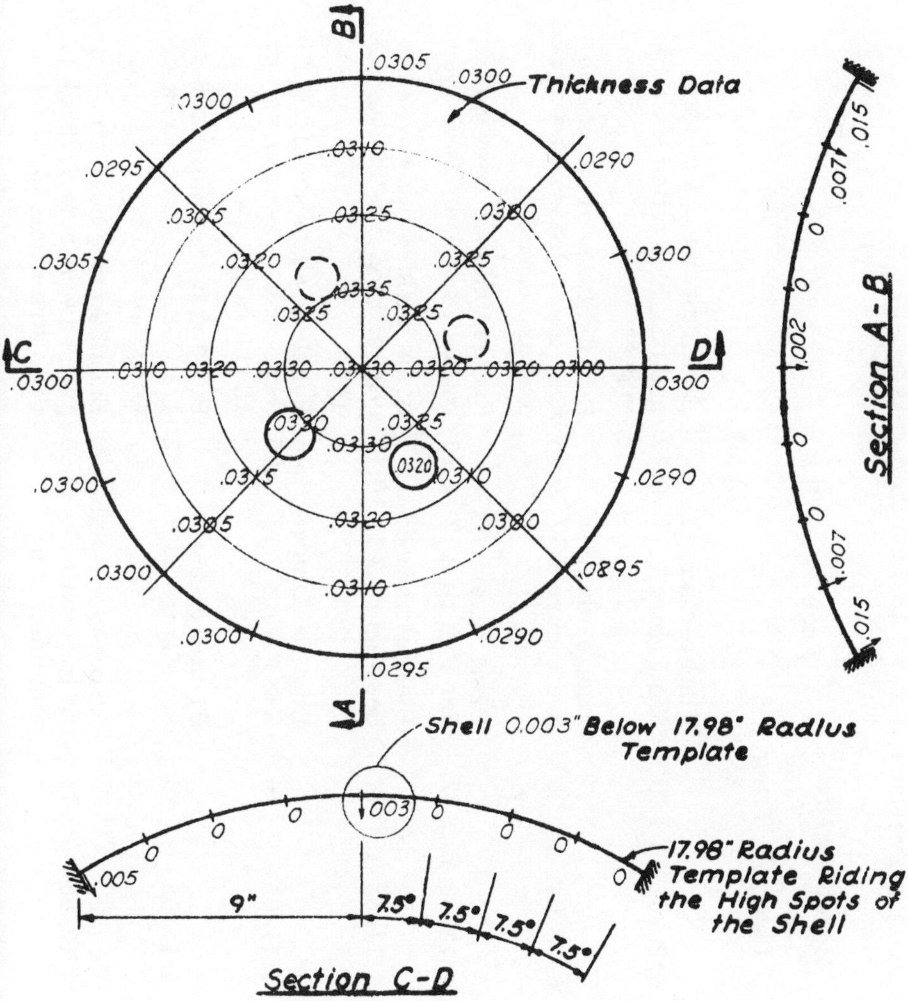

Maximum Thickness = 0.0335"
Minimum Thickness = 0.0290
Average Thickness = 0.0310
Thickness @ Air Pressure Buckle Position = 0.0320

○ Air Pressure Buckle Position

◌ Air Pressure Buckle Positions of Shells 6-1 and 6-2

Data Sheet for Shell 6-3

Test Data for Air-Pressure-Loaded 18-Inch Radius Domes

Shell	Modulus Test Temp. (°F)	Bending Modulus, Average from Four Samples (psi)	Model Test Date	Model Test Temp. (°F)	Edge Support	p_{cr} (psi)	Comments
1-1	80	449,000	6/20/62	70	Flange-clamped See Figure 2.17	0.49	Buckled twice. Pressure and buckle position near top repeat.
			6/20/62	70	Flange-clamped See Figure 2.17	0.53	Buckled twice at same spot near top. Pressures were 0.52 and 0.55 psi.
			4/11/63	72	Flange-clamped See Figure 2.17	0.53	Buckled midway between edge and top, not at 6/20/62 place.
			4/11/63	74	Flange-clamped See Figure 2.17	0.48	Again buckled midway between edge and top but 180 degrees away.
			4/11/63	75	Flange-clamped See Figure 2.17	0.55	Buckle position indifferent in three tests. Pressures were all the same.
			4/11/63	76	Flange-clamped See Figure 2.17	0.52	Buckled twice same place as 4/11/63, 74° test. Pressures were same.
			4/12/63	75	Epoxy-cemented up level with aluminum ring	0.60	Buckled twice near top 45 degrees away from 6/20/62 tests. Pressure same.

1-1	80	449,000	4/13/63	74	Epoxy-cemented only enough to seal the edge	0.60	Buckle position and pressure repeat in two tests.
1-2	80	469,000	6/21/62	70	Flange-clamped See Figure 2.17	0.51	Buckled twice. Pressures of 0.50 and 0.52 psi. Buckle position same.
			6/21/62	72	Flange-clamped See Figure 2.17	0.53	Buckled twice. Pressure repeats and buckle position same as before.
1-3	80	457,000	6/21/62	72	Flange-clamped See Figure 2.17	0.50	Buckled twice. Pressure and buckle position near top repeat.
			6/21/62	73	Flange-clamped See Figure 2.17	0.52	Buckled twice at same spot as before. Both pressures 0.52.
			7/7/62	76	Flange-clamped See Figure 2.17	0.53	Again buckled at same spot.
			7/7/62	79	Flange-clamped See Figure 2.17	0.55	Buckled twice at same spot as before. Pressures were 0.55 and 0.56.
			7/11/62	79	Epoxy-cemented up level with aluminum ring	0.51	Buckled five times at same spot as before. Pressures all the same.
2-1	72	446,000	7/12/62	81	Epoxy-cemented up level with aluminum ring	0.53	Buckled four times. Pressure and buckle position near top repeat.

Shell	Modulus Test Temp. (°F)	Bending Modulus, Average from Four Samples (psi)	Model Test Date	Model Test Temp. (°F)	Edge Support	p_{cr} (psi)	Comments
2-2	72	438,000	7/23/62	75	Epoxy-cemented up level with aluminum ring	0.56	Buckled four times. Pressure and buckle position near top repeat.
2-3	72	439,000	7/24/62	74	Epoxy-cemented up level with aluminum ring	0.62	Buckled four times. Buckle position near top repeats. Pressure from 0.61-0.62 psi.
3-1	83	439,000	7/15/62	72	Epoxy-cemented up level with aluminum ring	0.61	Buckled four times. Pressure and buckle position near top repeat.
			7/16/62	75	Epoxy-cemented up level with aluminum ring	0.61	Buckle position and pressure repeats in three tests.
3-2	83	454,000	7/25/62	74	Epoxy-cemented up level with aluminum ring	0.50	Buckled three times. Pressure and buckle position near top repeat.
3-3	83	448,000	7/17/62	73	Epoxy-cemented up level with aluminum ring	0.52	Buckled four times. Pressure and buckle position near top repeat.

3-4	83	446,000	7/26/62	76	Epoxy-cemented up level with aluminum ring	0.56	Buckled three times. Pressure and buckle position near top repeat.
3-5	83	450,000	7/18/62	72	Epoxy-cemented up level with aluminum ring	0.53	Buckled five times. Pressure and buckle position near top repeat.
4-1	74	449,000	7/14/62	78	Epoxy-cemented up level with aluminum ring	0.79	Buckled once at one place, three times at another. All near top and at same pressure.
4-2	78	444,000	7/20/62	78	Epoxy-cemented up level with aluminum ring	0.79	Buckled five times. Buckle position near top repeats, pressure from 0.76-0.82.
4-3	74	440,000	7/22/62	78	Epoxy-cemented up level with aluminum ring	0.83	Buckled five times. Pressure and buckle position near top repeat.
5-1	75	450,000	7/27/62	72	Epoxy-cemented up level with aluminum ring	0.98	Buckled four times. Pressure and buckle position near top repeat.
			3/11/63	77	Epoxy-cemented up level with aluminum ring	1.11	Buckled three times near top but 180° from 7/27/62 test. Pressure always 1.11.
			3/12/63	73	Epoxy-cemented up level with aluminum ring	1.11	Buckled three times at 7/27/62 position. Pressure repeats.

Shell	Modulus Test Temp. (°F)	Bending Modulus, Average from Four Samples (psi)	Model Test Date	Model Test Temp. (°F)	Edge Support	p_{cr} (psi)	Comments
			3/13/63	74	Epoxy-cemented up level with aluminum ring	1.11	Buckled three times at 3/11/63 position. Pressure repeats.
5-2	76	460,000	8/1/62	76	Epoxy-cemented up level with aluminum ring	0.76	Buckled four times. Pressure and buckle position near top repeat.
5-3	76	465,000	7/31/62	75	Epoxy-cemented up level with aluminum ring	0.77	Buckled four times. Buckle position near top repeats. Pressure from 0.76-0.77.
			9/7/62	66	Epoxy-cemented up level with aluminum ring	0.84	Buckled twice at the same position as before, pressure always 0.84.
			5/7/63	76	Epoxy-cemented up level with aluminum ring	0.77	Buckled three times at same position as before, pressure always 0.77.
6-1	80	470,000	6/23/63	74	Epoxy-cemented up level with aluminum ring	1.00	Buckled three times. Pressure and buckle position near top repeat.

			6/24/63	77	Epoxy-cemented up level with aluminum ring	1.01	Buckled two times at the same position as before, but at 1.01 psi pressure.
6-2	80	456,000	6/21/63	74	Epoxy-cemented up level with aluminum ring	1.00	Buckled two times. Pressure and buckle position near top repeat.
			6/25/63	81	Epoxy-cemented up level with aluminum ring	1.00	Buckled two times at the same pressure and position as before.
6-3	80	459,000	6/20/63	80	Epoxy-cemented up level with aluminum ring	1.02	Buckled three times. Pressure and buckle position near top repeat.
			6/26/63	82	Epoxy-cemented up level with aluminum ring	1.02	Buckled two times at same pressure as before, but 90° around the shell.
			6/27/63	84	Epoxy-cemented up level with aluminum ring	1.01	Buckled once at the same pressure and position as 6/20/63 test.

APPENDIX C

Data for Weight-Loaded 18-Inch Radius Shells

Weight Data for 20 Samples of Each Weight Used in Dead Weight Tests

May, 1963

	238.8 grams	119.5 grams	62.0 grams	11.6 gram
	238.9	118.8	61.4	11.6
	238.0	120.0	61.8	11.6
	239.0	119.4	61.0	11.6
	237.8	119.9	60.7	11.6
	239.8	119.5	61.4	11.7
	238.8	119.1	61.5	11.8
	237.4	119.2	61.9	11.7
	238.8	119.3	61.3	12.0
	238.1	119.2	62.2	12.3
	238.3	119.4	61.4	11.9
	239.1	119.9	61.4	11.6
	238.4	120.5	61.9	11.6
	238.3	121.2	61.5	11.8
	237.5	120.3	62.2	11.7
	238.6	119.4	62.5	11.7
	236.4	119.8	62.1	11.6
	239.6	120.0	61.7	12.3
	235.9	119.7	61.4	11.6
	239.0	121.2	60.7	11.6
High	= 239.8 grams	121.2 grams	62.5 grams	12.3 gram
Low	= 235.9 grams	118.7 grams	60.7 grams	11.6 gram
Average	= 238.3 grams	119.8 grams	61.6 grams	11.7 gram
	= 0.525 pound	= 0.264 pound	= 0.136 pound	= 0.026 p

Test Data for Weight-Loading Tests, 18-Inch Radius Domes

Shell	Air Pressure p_{cr}, psi	Model Test Date	Model Test Temp., °F	Edge Support	Loading Grid	Weight Loading p_{cr}, psi	Comments
1-1	0.570	5/7/63	84	Epoxy joint smaller than usual and did not cure properly	1-inch surface grid with loading pads	0.400	Buckled at thinnest place around the edge.
				Removed exposed portion of epoxy fillet	1-inch surface grid with loading pads	0.264	Buckled at same edge position as before.
		5/9/63	84-88	Epoxy joint up level with aluminum ring	1-inch surface grid with loading pads	0.536	Buckled at same edge position as before.
					1-inch surface grid without loading pads	0.536	Buckled at same edge position as before.
					2-inch surface grid without loading pads	0.562	Shell surface grossly distorted due to local bending.
1-3	0.522	4/9/63	–	Epoxy joint up level with aluminum ring	1-inch surface grid with loading pads	0.588	Buckled at thinnest place around the edge.

Shell	Air Pressure p_{cr}, psi	Model Test Date	Model Test Temp., °F	Edge Support	Loading Grid	Weight Loading p_{cr}, psi	Comments
3-1	0.610	3/31/63	–	Epoxy joint up level with aluminum ring	1-inch surface grid with loading pads	0.606	Buckled at thinnest place around the edge.
4-1	0.792	3/25/62	–	Epoxy joint up level with aluminum ring	1-inch surface grid with loading pads	0.661	Buckled at thinnest place around the edge. Removal of 0.136# from one string in center of buckle position permitted the addition of 70# distributed over the central portion of the shell.
					2-inch surface grid with loading pads	0.663	Shell surface grossly distorted due to local bending.
	0.862	4/5/63	77	Epoxy joint up level with aluminum ring	Air pressure		The string holes filled with plasticene clay and shell retested under air pressure.

					Buckle position near top as in previous air pressure tests.
0.897	4/7/63	79	Epoxy joint up level with aluminum ring	Air pressure	Buckle position near top as before.

APPENDIX D

Data for Air-Pressure-Loaded 36-Inch Radius Shells

Series 101. 36-Inch Radius Spherical Domes

Mold: Male plaster mold homemade. Plaster type was Ultracal 30. In order to make it lighter, the mold was filled with pieces of semirigid urethane foam plastic.

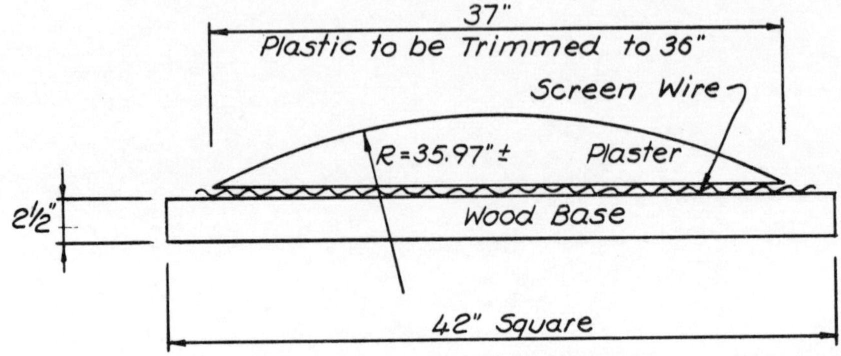

Plastic: Boltaron 6200 PVC purchased June 24, 1963.

Vacuum-Forming: Two models formed June 26, 1963, by the Gregstrom Corp., Cambridge, Massachusetts. I witnessed the forming and noted that the mold cracked at several places around the edge on the first model. Far too much heat was applied to the first model and still too much to the second.

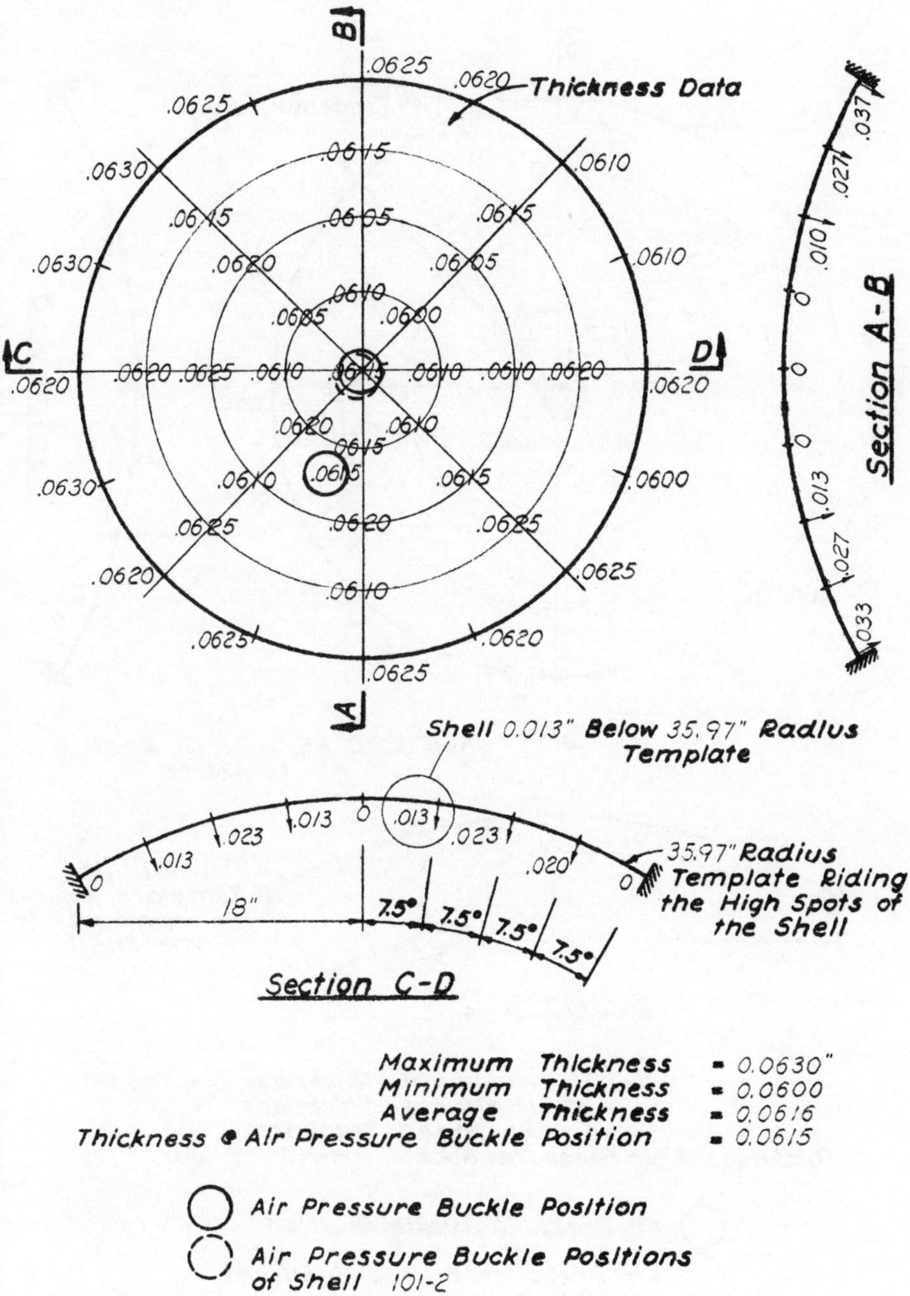

Data Sheet for Shell 101-1

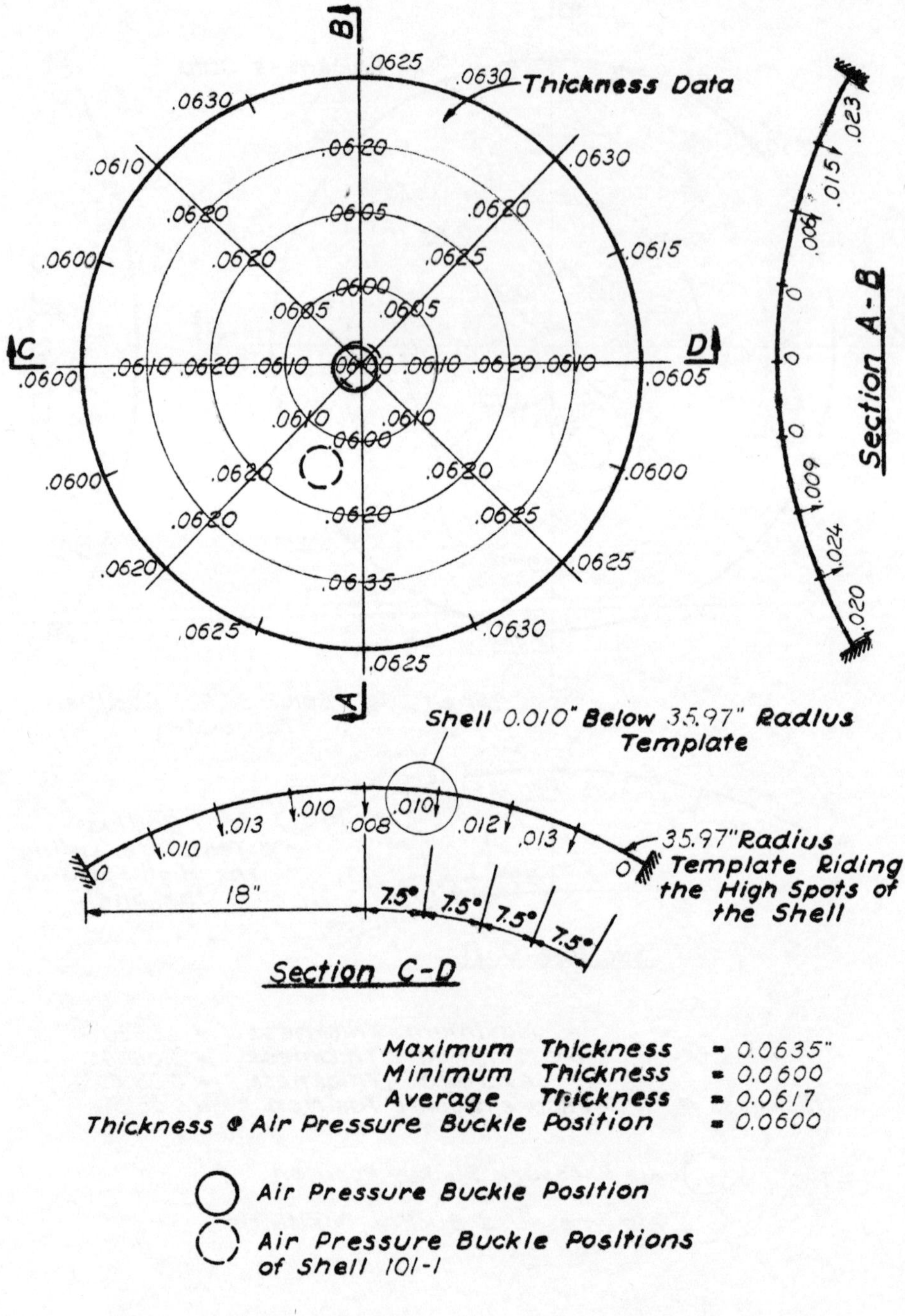

Data Sheet for Shell 101-2

Series 102. 36-Inch Radius Spherical Domes

Mold: Male plaster mold homemade. The mold was of solid Ultracal 30 plaster.

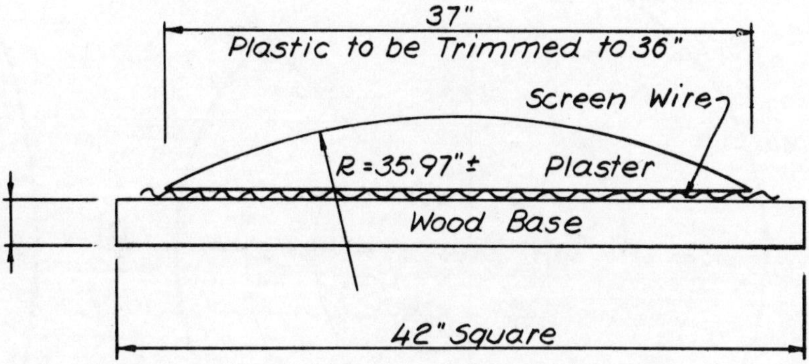

Plastic: Boltaron 6200 PVC purchased June 24, 1963.

Vacuum-Forming: Two models formed July 17, 1963, by the Gregstrom Corp., Cambridge, Massachusetts. I witnessed the forming and there were no difficulties.

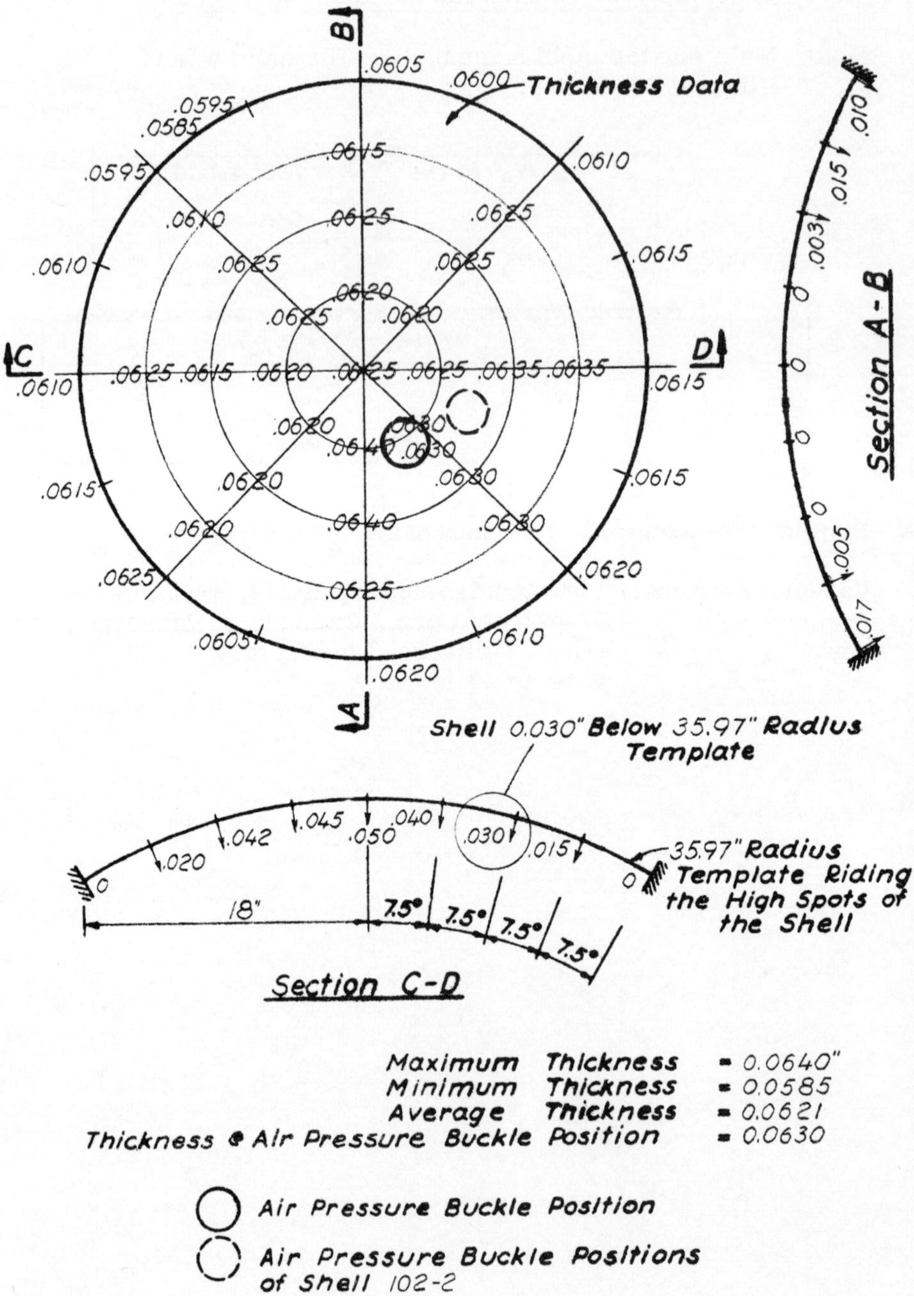

Data Sheet for Shell 102-1

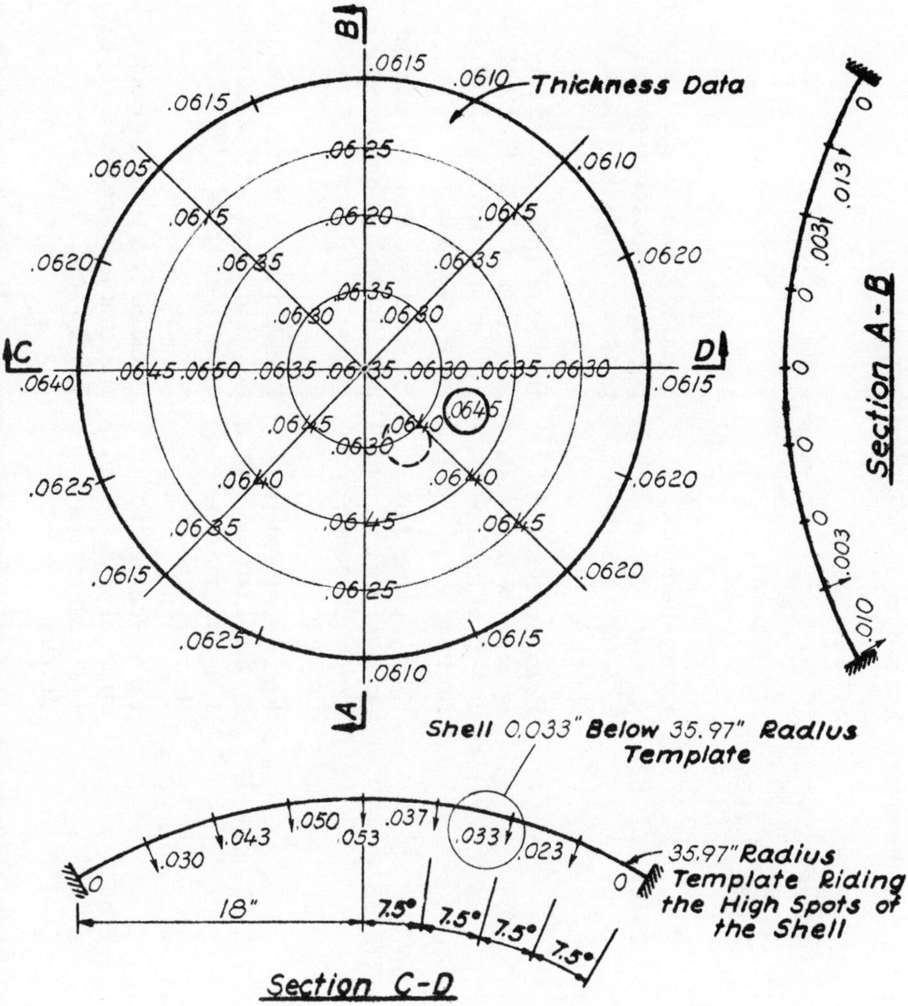

Maximum Thickness = 0.0650"
Minimum Thickness = 0.0605
Average Thickness = 0.0629
Thickness @ Air Pressure Buckle Position = 0.0645

○ Air Pressure Buckle Position

◌ Air Pressure Buckle Positions of Shell 102-1

Data Sheet for Shell 102-2

Test Data for Air-Pressure-Loaded 36-Inch Radius Domes

Shell	Modulus Test Temp. (°F)	Bending Modulus, Average from Four Samples (psi)	Model Test Date	Model Test Temp. (°F)	Edge Support	p_{cr} (psi)	Comments
101-1	86	403,000	6/28/63	82	Epoxy-cemented up level with 1/4-inch mason-ite ring	0.85	Buckled three times. Buckle position near top repeats, pressure from 0.84-0.85.
			7/4/63	81	Epoxy-cemented up level with 1/4-inch mason-ite ring	0.90	Buckle position at top for first test at 6/28/63 position for 2nd and 3rd tests. Pres-sure always 0.90.
101-2	85	410,000	7/2/63	91	Epoxy-cemented up level with 1/4-inch mason-ite ring	0.71	Buckled two times. Buckle position at top and pressure repeat.
			7/5/63	75	Epoxy-cemented up level with 1/4-inch mason-ite ring	0.78	Buckled three times. Buckle position same as 7/2/63 test, pressures from 0.77-0.79.
			7/7/63	74	Epoxy-cemented up level with 1/4-inch mason-ite ring	0.78	Buckled two times. Buckle position and pressure repeat 7/5/63 test.

102-1	85	411,000	7/18/63	85	Epoxy-cemented up level with 1/4-inch mason-ite ring	0.98	Buckled four times. Buckle position repeats. Pressure varies 0.96-0.99 psi.
102-2	85	415,000	7/20/63	83	Epoxy-cemented up level with 1/4-inch mason-ite ring	1.02	Buckled three times. Buckle position repeats. Pressure varies 1.02-1.03 psi.

REFERENCES

1. Axilrod, Sherman, Cohen, and Wolock, "Effects of Biaxial Stretch Forming," Mod. Plastics, 30, 117 (December, 1952).
2. Bailey, J., "Stretch Orientation of Polystyrene and Its Interesting Results," India Rubber World, 118, 225 (1948).
3. Beaujoint, N., "Similitude and Theory of Models," R. I. L. E. M. Bulletin No. 7 (June, 1960). This article deals exclusively with structural models.
4. Beers, Y., Introduction to the Theory of Error, Addison-Wesley Publishing Company, Inc., Reading, Mass. (1957). This book (and Parratt's) provides an introduction to the treatment of random errors. They attempt to show how the theory of probability and statistics can be used in an experimental study rather than in classical mathematical exercises.
5. Much of the information regarding the properties of the Boltaron plastics, and in particular the information contained in Table 2.2, and Figures 2.10 and 2.13, was obtained directly from the manufacturer or from Bolta Products Sales Bulletin S-61-5 (1961).
6. Bridgman, P. W., Dimensional Analysis, Yale University Press, New Haven (1922). Bridgman, a physicist, clarified many of the problems of the day, and his work has become a classic in the field.
7. Bryan, G. H., "On the Stability of Elastic Systems," Proc. Cambridge Phil. Soc., 6, 199 (1888).
8. Buckingham, E., "On Physically Similar Systems," Phys. Rev., 4, 345 (1914). Buckingham's work is one of the milestones in the exposition of the theory of dimensions.
9. Donnell, L. H., "A New Theory for the Buckling of Thin Cylinders under Axial Compression and Bending," Trans. Am. Soc. Mech. Engrs., 56, 795 (1934).
10. Dove, R. C., "Strain Measurement Errors in Materials of Low Modulus," Proc. Am. Soc. Civil Engrs., Eng. Mech. Div., 81, Separate 691 (May, 1955).

11. Euler, L., "Elastic Curves," Isis, 20, 72 (November, 1933), English translation.

12. Fairbairn, W., "On the Resistance of Tubes to Collapse," Proc. Roy. Soc. London, 9, 234 (1858).

13. Findley, N. N., "Creep of Plastics—Mechanism and Mechanics," Soc. Plastics Engrs. J., 16, 57 and 192 (1960).

14. Flügge, W., Stresses in Shells, Springer-Verlag, Berlin, (1960).

15. Fung, Y. C., and Sechler, E. E., "Instability of Thin Elastic Shells," Proceedings of First Symposium on Naval Structural Mechanics, edited by Goodier and Hoff, Stanford University Press, Stanford, California (1958).

16. Gerard, G., and Becker, H., "Handbook of Structural Stability; Part III: Buckling of Curved Plates and Shells," NACA Tech. Note 3783 (1957).

17. Hansen, R. J., and Litle, W. A., M. I. T. class notes (to be published).

18. Hognestad, E., Hanson, N. W., and McHenry, D., "Concrete Stress Distribution in Ultimate Strength Design," J. Am. Concrete Inst., 52, 455 (1955).

19. Homewood, R. H., Brine, A. C., and Johnston, A. E., "Experimental Investigation of the Buckling Instability of Monocoque Shells," Proc. Soc. Exptl. Stress Anal., 18, No. 1, 88 (1961).

20. Johnson, A. E., Jr., and Homewood, R. H., "Stress and Deformation Analysis from Reduced Scale Plastic Model Testing," Proc. Soc. Exptl. Stress Anal., 18, No. 2, 81 (1961).

21. Kaplan, A., and Fung, Y. C., "A Nonlinear Theory of Bending and Buckling of Thin Elastic Shallow Spherical Shells," NACA Tech. Note 3212 (April, 1954).

22. von Karman, T., and Tsien, H. S., "The Buckling of Spherical Shells by External Pressure," J. Aeron. Sci. 7, 43 (December, 1939). See also von Karman, T., and Tsien, H. S., "The Buckling of Thin Cylindrical Sheels under Axial Compression," J. Aeron. Sci., 8, 303 (June, 1941).

23. Kinney, G. F., Engineering Properties and Applications of Plastics, John Wiley & Sons, Inc., New York (1957), p. 5.

24. Klöppel, K., and Jungbluth, O., "Beitrag zum Durchschagproblem dünnwandiger Kugelschalen," Der Stahlbau (June, 1953).

25. Langhaar, H. L., Dimensional Analysis and Theory of Models, John Wiley & Sons, Inc., New York (1951). Murphy and Langhaar have written two of the many books dealing with dimensional analysis. In both books, the examples are drawn from the hydraulic, thermal, electrical, and structural fields, Murphy presenting more of a practical approach, and Langhaar being somewhat more theoretical.

26. Lundquist, E. E., "Strength Tests of Thin-Walled Duralumin Cylinders in Compression," NACA Rept. 473 (1933).

27. Manufacturing Chemist's Association, Washington, D. C., "Technical Data on Plastics" (1957).

28. Maxwell, B., and Harrington, J. P., "Effect of Velocity on Tensile Impact Properties of Polymethyl Methacrylate," Trans. Am. Soc. Mech. Engrs., 74, 579 (1952).

29. Murphy, G., Similitude in Engineering, Ronald Press, New York (1950).

30. Nash, W. A., "Bibliography on Shells and Shell-like Structures 1954-1956," University of Florida, Contract DA-01009 -ORD-404, Office of Ordnance Research, U. S. Army (1957).

31. Nash, W. A., "Bibliography on Shells and Shell-like Structures," TBM Report 863, U. S. Navy Department, David Taylor Model Basin, Washington, D. C. (1954).

32. Northrop Aircraft, Inc., "Mechanical Properties of Formed Methyl Methyacrylate," Report No. L. N. 2376 (April 14, 1948).

33. O'Conner, D. G., and Findley, N. N., "Influence of Normal Stress on Creep in Tension and Compression of Polyethylene and Rigid Polyvinyl Chloride Copolymer," Trans. Am. Soc. Mech. Engrs., Ser. B, Paper No. 61-WA-35 (1961).

34. Pahl, P. J., "A General Theory of Physical Models and Its Application to Structures with Significant Mass," Department of Civil Engineering, M. I. T., Cambridge, Mass., T62-6 (1962).

35. Pahl, P. J., "Confidence Levels for Structural Models," Department of Civil Engineering, M. I. T., Cambridge, Mass., T63-5 (February, 1963).

36. Parratt, L. G., Probability and Experimental Errors in Science, John Wiley & Sons, Inc., New York (1961).

37. Parzen, E., Modern Probability Theory and Its Application, John Wiley & Sons, Inc., New York (1960). This textbook covers the general introduction to probability theory. The

layman may not appreciate the mathematical preciseness.

38. Schmidt, L., "Ergebnisse von Beulversuchen mit doppelt gekrümmten Schalenmodellen aus Aluminum," Proceedings of a Symposium on Shell Research, Delft, the Netherlands, North-Holland Publishing Company, Amsterdam (1961).

39. Seaman, L., "The Nature of Buckling in Thin Spherical Shells," Ph. D. Thesis, Department of Civil Engineering, M. I. T., Cambridge, Mass. (1961).

40. Soosaar, K., "Systematic Errors in the Loading of Stress Models of Shells," Department of Civil Engineering, M. I. T., Cambridge, Mass., T63-6 (May, 1963).

41. Southwell, R. V., "On the General Theory of Elastic Stability," Phil. Trans. Roy. Soc. London, Ser. A, 213, 187 (1914).

42. Sturm, R. G., "A Study of the Collapsing Pressure of Instability of Thin-Walled Cylinders," Bulletin No. 329, University of Illinois Engineering Experimental Station (November, 1941).

43. Timoshenko, S., and Gere, J., Theory of Elastic Stability, second edition, McGraw-Hill Book Company, Inc., New York (1961).

44. Tsien, H. S., "A Theory for the Buckling of Thin Shells," J. Aeron. Sci., 9, 373 (1942).

45. Weinitschke, H. J., "Asymmetric Buckling of Clamped Shallow Spherical Shells," NASA Tech. Note D-1510 (1962); Huang, N. C., "Unsymmetrical Buckling of Thin Shallow Spherical Shells," Harvard University Technical Report No. 15 (March, 1963); Krenzke, M. A., and Kiernan, T. J., "Elastic Stability of Near-Perfect Shallow Spherical Shells," J. Am. Inst. Aeronautics and Astronautics, 2855-2857 (December, 1963).

46. Wilson, E. B., Jr., An Introduction to Scientific Research, McGraw-Hill Book Company, Inc., New York (1952). A book which brings together many of the too often forgotten general principles, techniques, and guides for procedure.

47. Zoelly, R., Dissertation, Zurich (1915).

INDEX